Sectional Anatomy Review

Sectional Anatomy Review

Michael E. Madden, PhD, RT(R), (CT), (MR)

Professor

Department of Allied Health

Fort Hays State University

LIPPINCOTT WILLIAMS & WILKINS
A **Wolters Kluwer** Company

Editor: Lawrence McGrew
Associate Managing Editor: Angela Heubeck
Marketing Manager: Debby Hartman
Production Editor: Jennifer D. Weir

Copyright © 2001 Lippincott Williams & Wilkins

530 Walnut Street
Philadelphia, Pennsylvania 19106

351 West Camden Street
Baltimore, Maryland 21201-2436 USA

The publisher is not responsible (as a matter of product liability, negligence, or otherwise) for any injury resulting from any material contained herein. This publication contains information relating to general principles of medical care which should not be construed as specific instructions for individual patients. Manufacturers' product information and package inserts should be reviewed for current information, including contraindications, dosages and precautions.

Printed in the United States of America

The publishers have made every effort to trace the copyright holders for borrowed material. If they have inadvertently overlooked any, they will be pleased to make the necessary arrangements at the first opportunity.

To purchase additional copies of this book, call our customer service department at **(800) 638-3030** or fax orders to **(301) 824-7390.** For other book services, including chapter reprints and large quantity sales, ask for the Special Sales department.

For all other calls originating outside of the United States, please call **(301)714-2324.**

Visit Lippincott Williams & Wilkins on the Internet: **http://www.lww.com**. Lippincott Williams & Wilkins customer service representatives are available from 8:30 am to 6:00 pm, EST, Monday through Friday, for telephone access.

00 01 02 03 04
1 2 3 4 5 6 7 8 9 10

Contents

Chapter 1

Chest

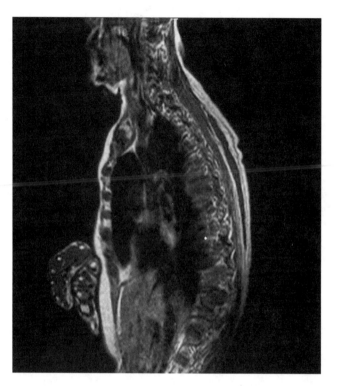 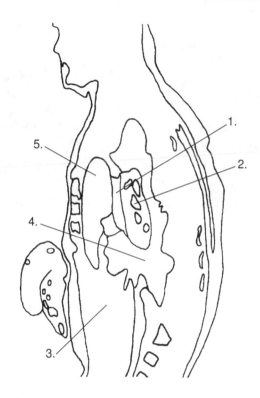

Figure 1-1

1. Which of the following is illustrated by 4?
    _____ A. Liver
    _____ B. Right lung lower lobe
    _____ C. Left lung lower lobe
    _____ D. Hilar region
2. Which of the following is illustrated by 2?
    _____ A. Aortic arch
    _____ B. Superior vena cava
    _____ C. Hilar region
    _____ D. Heart
3. What number illustrates the upper lobe of the lung?
    _____ A. 3
    _____ B. 1
    _____ C. 4
    _____ D. 5

4. Which of the following is illustrated by 3?
    _____ A. Stomach
    _____ B. Liver
    _____ C. Right lung lower lobe
    _____ D. Left lung lower lobe
5. What number illustrates the superior vena cava?
    _____ A. 1
    _____ B. 5
    _____ C. 4
    _____ D. 2

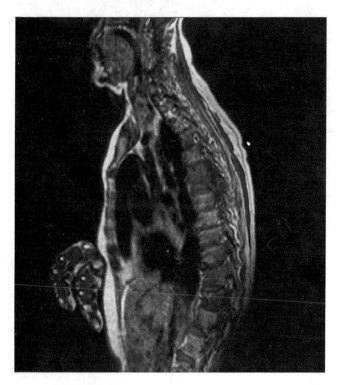
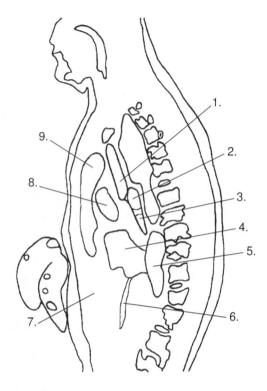

Figure 1-2

1. Which of the following is illustrated by 2?

    _____ A. Superior vena cava

    _____ B. Right pulmonary artery

    _____ C. Right pulmonary vein

    _____ D. Wall of the aortic arch

2. What number illustrates the inferior vena cava?

    _____ A. 5

    _____ B. 4

    _____ C. 3

    _____ D. 6

3. Which of the following is illustrated by 3?

    _____ A. Right pulmonary artery

    _____ B. Right pulmonary vein

    _____ C. Right atrium

    _____ D. Superior vena cava

4. What number illustrates the wall of aortic arch?

    _____ A. 9

    _____ B. 7

    _____ C. 8

    _____ D. 1

5. Which of the following is illustrated by 4?

    _____ A. Liver

    _____ B. Right atrium

    _____ C. Right lung lower lobe

    _____ D. Inferior vena cava

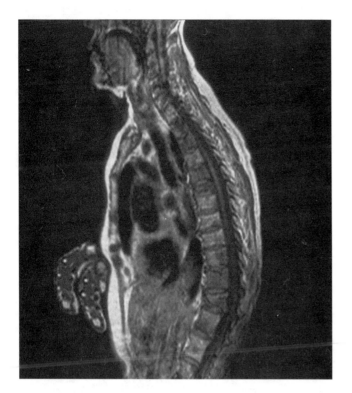
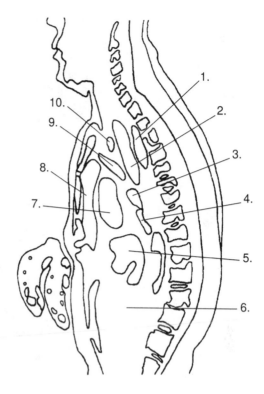

Figure 1-3

1. Which of the following is illustrated by 2?

_____ A. Esophagus

_____ B. Brachiocephalic artery

_____ C. Trachea

_____ D. Left brachiocephalic vein

2. What number illustrates the right lung upper lobe?

_____ A. 7

_____ B. 3

_____ C. 8

_____ D. 9

3. Which of the following is illustrated by 10?

_____ A. Left brachiocephalic vein

_____ B. Right brachiocephalic vein

_____ C. Brachiocephalic artery

_____ D. Left subclavian artery

4. Which of the following is illustrated by 4?

_____ A. Right pulmonary artery

_____ B. Left atrium

_____ C. Right atrium

_____ D. Left ventricle

5. What number illustrates the left brachiocephalic vein?

_____ A. 10

_____ B. 2

_____ C. 9

_____ D. 1

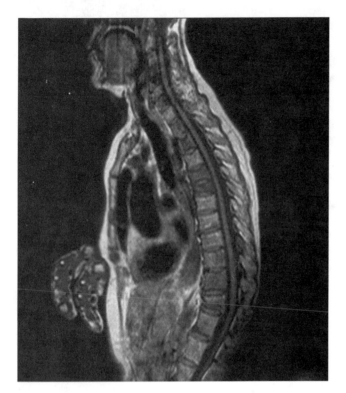
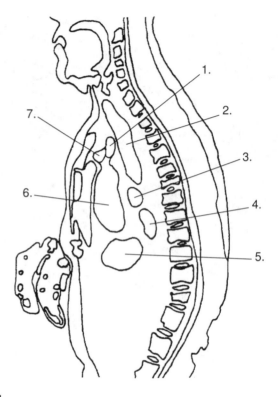

Figure 1-4

1. Which of the following is illustrated by 7?

   _____ A. Brachiocephalic artery

   _____ B. Left brachiocephalic vein

   _____ C. Aortic arch

   _____ D. Pulmonary artery

2. Which of the following is illustrated by 6?

   _____ A. Right pulmonary artery

   _____ B. Left atrium

   _____ C. Aortic arch

   _____ D. Left brachiocephalic vein

3. What number illustrates the brachiocephalic artery?

   _____ A. 1

   _____ B. 7

   _____ C. 3

   _____ D. 2

4. Which of the following is illustrated by 5?

   _____ A. Right pulmonary artery

   _____ B. Left atrium

   _____ C. Right atrium

   _____ D. Aortic arch

5. What number illustrates the right pulmonary artery?

   _____ A. 3

   _____ B. 6

   _____ C. 5

   _____ D. 4

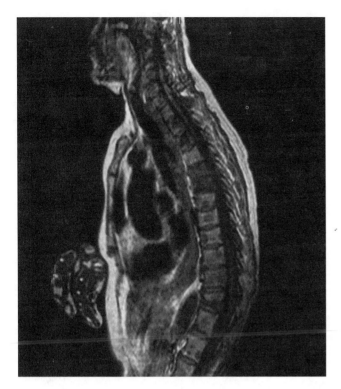 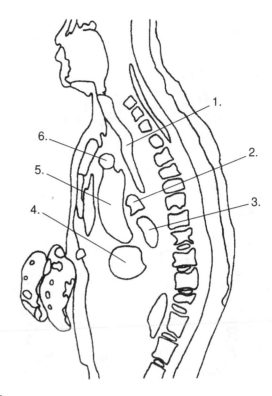

Figure 1-5

1. What number illustrates the right ventricle?

　_____ A. 6

　_____ B. 3

　_____ C. 4

　_____ D. 2

2. Which of the following is illustrated by 6?

　_____ A. Right pulmonary artery

　_____ B. Brachiocephalic artery

　_____ C. Aortic arch

　_____ D. Left brachiocephalic vein

3. Which of the following is illustrated by 5?

　_____ A. Left atrium

　_____ B. Right atrium

　_____ C. Aortic arch

　_____ D. Right ventricle

4. What number illustrates the right pulmonary artery?

　_____ A. 2

　_____ B. 6

　_____ C. 5

　_____ D. 3

5. Which of the following is illustrated by 3?

　_____ A. Right ventricle

　_____ B. Left atrium

　_____ C. Aortic arch

　_____ D. Right atrium

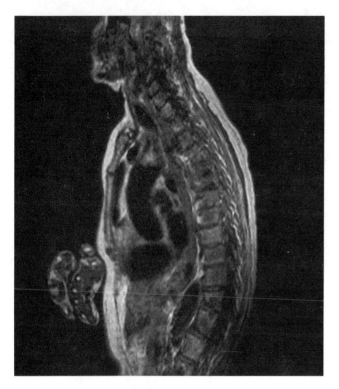
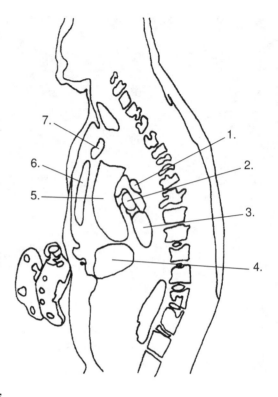

Figure 1-6

1. Which of the following is illustrated by 5?

   _____ A. Pleural cavity

   _____ B. Right atrium

   _____ C. Aortic arch

   _____ D. Left atrium

2. What number illustrates the left bronchus?

   _____ A. 1

   _____ B. 2

   _____ C. 5

   _____ D. 3

3. What number illustrates the pleural cavity?

   _____ A. 3

   _____ B. 4

   _____ C. 6

   _____ D. 5

4. Which of the following is illustrated by 2?

   _____ A. Left bronchus

   _____ B. Right pulmonary artery

   _____ C. Left atrium

   _____ D. Aortic arch

5. Which of the following is illustrated by 3?

   _____ A. Pleural cavity

   _____ B. Right ventricle

   _____ C. Left atrium

   _____ D. Aortic arch

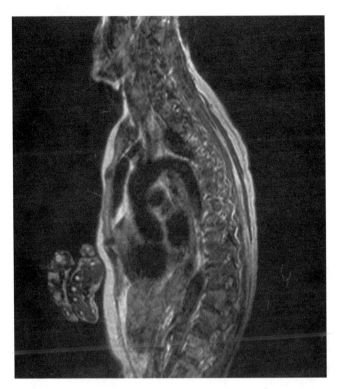 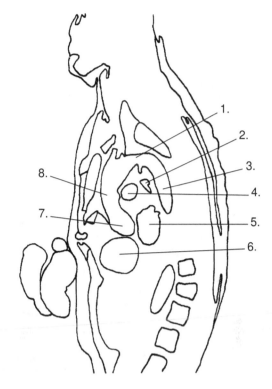

Figure 1-7

1. Which of the following is illustrated by 3?

    _____ A. Aortic arch

    _____ B. Left bronchus

    _____ C. Descending aorta

    _____ D. Pulmonary artery

2. What number illustrates the aortic arch?

    _____ A. 1

    _____ B. 4

    _____ C. 3

    _____ D. 2

3. Which of the following is illustrated by 7?

    _____ A. Left atrium

    _____ B. Right ventricle

    _____ C. Left ventricle

    _____ D. Ascending aorta

4. Which of the following is illustrated by 4?

    _____ A. Left bronchus

    _____ B. Descending aorta

    _____ C. Pulmonary artery

    _____ D. Left atrium

5. What number illustrates the ascending aorta?

    _____ A. 7

    _____ B. 5

    _____ C. 4

    _____ D. 8

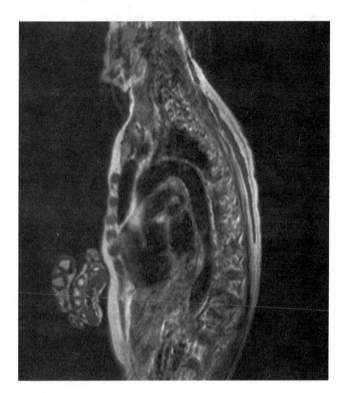
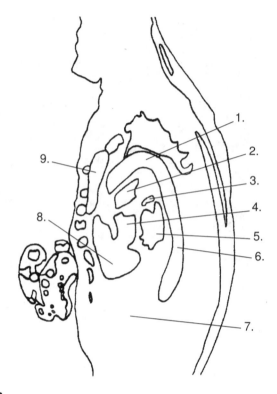

Figure 1-8

1. Which of the following is illustrated by 2?

_____ A. Aortic arch

_____ B. Pulmonary trunk

_____ C. Right atrium

_____ D. Left atrium

2. What number illustrates the right ventricle?

_____ A. 2

_____ B. 4

_____ C. 8

_____ D. 5

3. Which of the following is illustrated by 5?

_____ A. Aortic arch

_____ B. Left ventricle

_____ C. Right ventricle

_____ D. Left atrium

4. Which of the following is illustrated by 3?

_____ A. Left bronchus

_____ B. Left ventricle

_____ C. Left atrium

_____ D. Pulmonary trunk

5. What number illustrates the left ventricle?

_____ A. 2

_____ B. 5

_____ C. 4

_____ D. 8

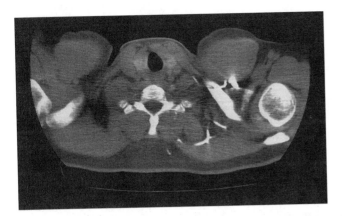 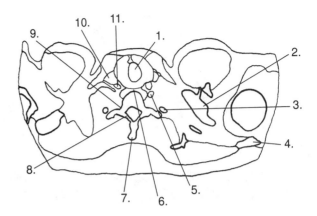

Figure 1-9

1. What number illustrates the right common carotid artery?
   _____ A. 10
   _____ B. 5
   _____ C. 3
   _____ D. 11

2. Which of the following is illustrated by 10?
   _____ A. Right external jugular vein
   _____ B. Right internal jugular vein
   _____ C. Right common carotid artery
   _____ D. Right subclavian artery

3. What number illustrates the clavicle?
   _____ A. 2
   _____ B. 4
   _____ C. 1
   _____ D. 8

4. Which of the following is illustrated by 9?
   _____ A. Spinous process of T1
   _____ B. Pedicle of T1
   _____ C. Transverse process of T1
   _____ D. Lamina of T1

5. Which of the following is illustrated by 6?
   _____ A. Thyroid
   _____ B. Trachea
   _____ C. Esophagus
   _____ D. Larynx

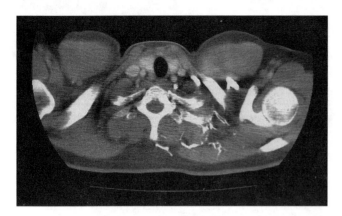 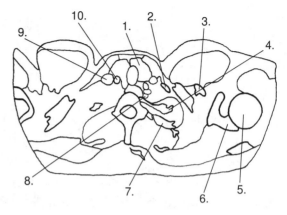

Figure 1-10

1. Which of the following is illustrated by 9?

    _____ A. Right internal jugular vein

    _____ B. Left subclavian vein

    _____ C. Left vertebral artery

    _____ D. Right common carotid artery

2. What number illustrates the first rib?

    _____ A. 6

    _____ B. 4

    _____ C. 7

    _____ D. 3

3. Which of the following is illustrated by 8?

    _____ A. Left subclavian vein

    _____ B. Left subclavian artery

    _____ C. Left vertebral artery

    _____ D. Left axillary vein

4. Which of the following is illustrated by 7?

    _____ A. Glenoid process of scapula

    _____ B. First rib

    _____ C. Second rib

    _____ D. Head of the humerus

5. What number illustrates the right common carotid artery?

    _____ A. 9

    _____ B. 8

    _____ C. 2

    _____ D. 10

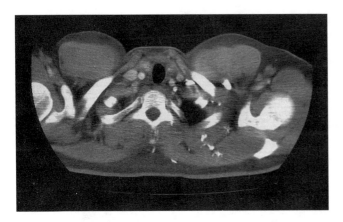
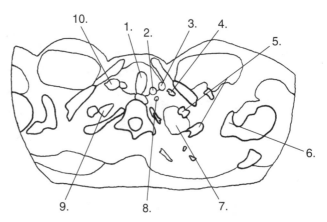

Figure 1-11

1. Which of the following is illustrated by 9?

    _____ A. Axillary vein

    _____ B. Upper lobe left lung

    _____ C. Upper lobe of right lung

    _____ D. Internal jugular vein

2. What number illustrates the coracoid process of scapula?

    _____ A. 9

    _____ B. 6

    _____ C. 5

    _____ D. 1

3. What number illustrates the left vertebral artery?

    _____ A. 2

    _____ B. 8

    _____ C. 4

    _____ D. 3

4. Which of the following is illustrated by 4?

    _____ A. Left internal jugular vein

    _____ B. Left common carotid artery

    _____ C. Left axillary artery

    _____ D. Left subclavian vein

5. Which of the following is illustrated by 2?

    _____ A. Left common carotid artery

    _____ B. Trachea

    _____ C. Left subclavian vein

    _____ D. Left vertebral artery

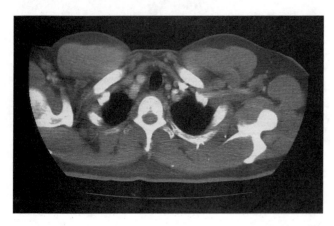 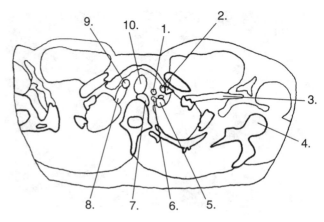

Figure 1-12

1. Which of the following is illustrated by 2?

   _____ A. Left common carotid artery

   _____ B. Left brachiocephalic vein

   _____ C. Left axillary artery

   _____ D. Left vertebral artery

2. What number illustrates the left axillary artery?

   _____ A. 3

   _____ B. 7

   _____ C. 6

   _____ D. 5

3. Which of the following is illustrated by 6?

   _____ A. Left subclavian artery

   _____ B. Left subclavian vein

   _____ C. Left vertebral artery

   _____ D. Left common carotid artery

4. What number illustrates the right common carotid artery?

   _____ A. 9

   _____ B. 10

   _____ C. 1

   _____ D. 8

5. Which of the following is illustrated by 1?

   _____ A. Left brachiocephalic vein

   _____ B. Left vertebral artery

   _____ C. Left subclavian artery

   _____ D. Left common carotid artery

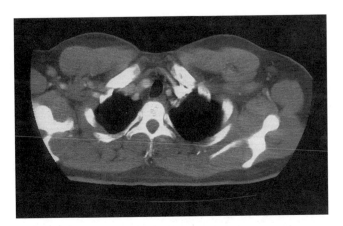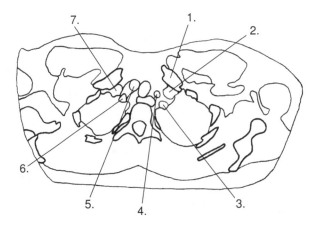

Figure 1-13

1. Which of the following is illustrated by 2?

   _____ A. Left clavicle

   _____ B. Left brachiocephalic vein

   _____ C. Left subclavian artery

   _____ D. Left common carotid artery

2. What number illustrates the origin of the right common carotid artery?

   _____ A. 7

   _____ B. 5

   _____ C. 6

   _____ D. 4

3. Which of the following is illustrated by 1?

   _____ A. Left clavicle

   _____ B. Left brachiocephalic vein

   _____ C. Left subclavian artery

   _____ D. Left common carotid artery

4. Which of the following is illustrated by 6?

   _____ A. Origin of the right vertebral artery

   _____ B. Origin of the right common carotid artery

   _____ C. Origin of the right subclavian artery

   _____ D. Right brachiocephalic vein

5. What number illustrates the left common carotid artery?

   _____ A. 2

   _____ B. 6

   _____ C. 4

   _____ D. 3

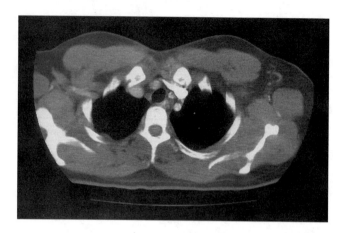
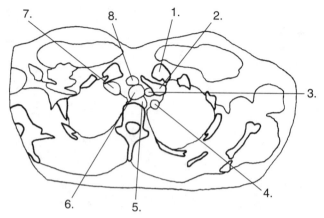

Figure 1-14

1. What number illustrates the brachiocephalic artery?
   _____ A. 7
   _____ B. 6
   _____ C. 4
   _____ D. 8

2. Which of the following is illustrated by 6?
   _____ A. Left subclavian artery
   _____ B. Esophagus
   _____ C. Trachea
   _____ D. Right brachiocephalic vein

3. Which of the following is illustrated by 4?
   _____ A. Left common carotid artery
   _____ B. Left subclavian artery
   _____ C. Left vertebral artery
   _____ D. Left axillary artery

4. What number illustrates the left brachiocephalic vein?
   _____ A. 2
   _____ B. 8
   _____ C. 1
   _____ D. 4

5. Which of the following is illustrated by 7?
   _____ A. Left brachiocephalic vein
   _____ B. Right subclavian vein
   _____ C. Brachiocephalic artery
   _____ D. Right brachiocephalic vein

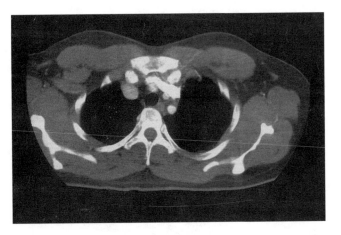 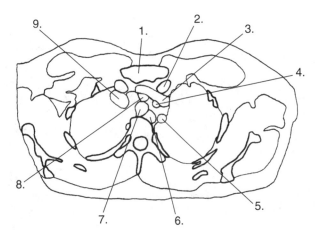

Figure 1-15

1. Which of the following is illustrated by 6?

   _____ A. Left subclavian artery

   _____ B. Esophagus

   _____ C. Trachea

   _____ D. Left common carotid artery

2. What number illustrates the right brachiocephalic vein?

   _____ A. 8

   _____ B. 7

   _____ C. 4

   _____ D. 9

3. Which of the following is illustrated by 1?

   _____ A. Clavicle

   _____ B. Trachea

   _____ C. Left brachiocephalic vein

   _____ D. Manubrium

4. Which of the following is illustrated by 4?

   _____ A. Left subclavian vein

   _____ B. Subclavian artery

   _____ C. Left common carotid artery

   _____ D. Left brachiocephalic vein

5. What number illustrates the brachiocephalic artery?

   _____ A. 9

   _____ B. 6

   _____ C. 8

   _____ D. 4

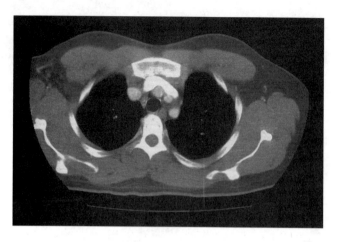
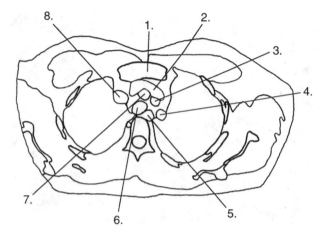

Figure 1-16

1. What number illustrates the left brachiocephalic vein?

    _____ A. 2

    _____ B. 8

    _____ C. 7

    _____ D. 4

2. Which of the following is illustrated by 5?

    _____ A. Left common carotid artery

    _____ B. Esophagus

    _____ C. Trachea

    _____ D. Manubrium

3. Which of the following is illustrated by 4?

    _____ A. Left vertebral artery

    _____ B. Brachiocephalic artery

    _____ C. Left subclavian artery

    _____ D. Left common carotid artery

4. What number illustrates the brachiocephalic artery?

    _____ A. 4

    _____ B. 2

    _____ C. 3

    _____ D. 7

5. Which of the following is illustrated by 6?

    _____ A. Right brachiocephalic vein

    _____ B. Esophagus

    _____ C. Trachea

    _____ D. Right internal jugular vein

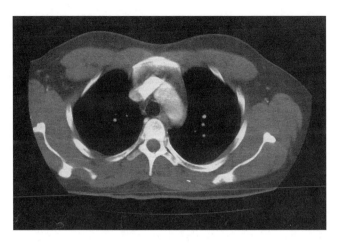 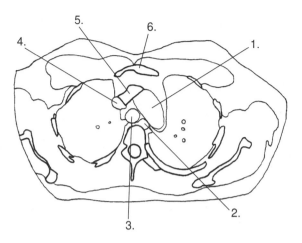

Figure 1-17

1. Which of the following is illustrated by 2?

    _____ A. Trachea

    _____ B. Vertebral artery

    _____ C. Aortic arch

    _____ D. Esophagus

2. What number illustrates the right brachiocephalic vein?

    _____ A. 5

    _____ B. 3

    _____ C. 4

    _____ D. 2

3. What number illustrates the aortic arch?

    _____ A. 1

    _____ B. 4

    _____ C. 5

    _____ D. 2

4. Which of the following is illustrated by 5?

    _____ A. Aortic arch

    _____ B. Left brachiocephalic vein

    _____ C. Brachiocephalic artery

    _____ D. Right brachiocephalic vein

5. Which of the following is illustrated by 3?

    _____ A. Aortic arch

    _____ B. Trachea

    _____ C. Esophagus

    _____ D. Left main bronchus

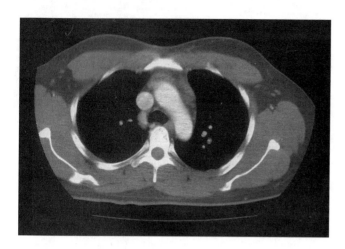 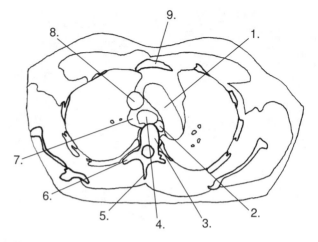

Figure 1-18

1. What number illustrates the azygos arch?

    _____ A. 7

    _____ B. 8

    _____ C. 2

    _____ D. 1

2. Which of the following is illustrated by 3?

    _____ A. Esophagus

    _____ B. Vertebral body

    _____ C. Trachea

    _____ D. Azygos arch

3. Which of the following is illustrated by 5?

    _____ A. Pedicle

    _____ B. Lamina

    _____ C. Spinous process

    _____ D. Transverse process

4. What number illustrates the transverse process?

    _____ A. 7

    _____ B. 5

    _____ C. 6

    _____ D. 1

5. Which of the following is illustrated by 8?

    _____ A. Inferior vena cava

    _____ B. Ascending aorta

    _____ C. Azygos arch

    _____ D. Superior vena cava

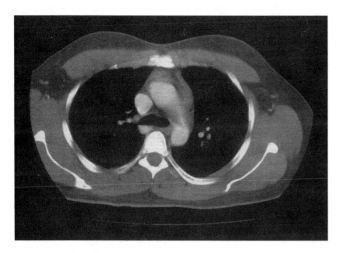 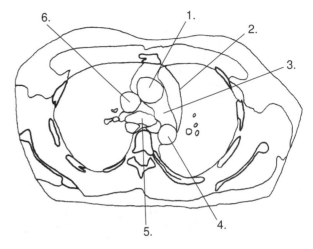

Figure 1-19

1. Which of the following is illustrated by 2?

_____ A. Right atrium

_____ B. Mediastinal pleura

_____ C. Left atrium

_____ D. Right ventricle

2. What number illustrates the tracheal bifurcation?

_____ A. 6

_____ B. 3

_____ C. 5

_____ D. 2

3. Which of the following is illustrated by 4?

_____ A. Mediastinal pleura

_____ B. Ascending aorta

_____ C. Bottom of the aortic arch

_____ D. Descending aorta

4. Which of the following is illustrated by 1?

_____ A. Ascending aorta

_____ B. Bottom of the aortic arch

_____ C. Superior vena cava

_____ D. Descending aorta

5. What number illustrates the bottom of the aortic arch?

_____ A. 5

_____ B. 3

_____ C. 4

_____ D. 1

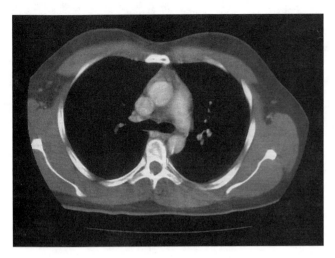 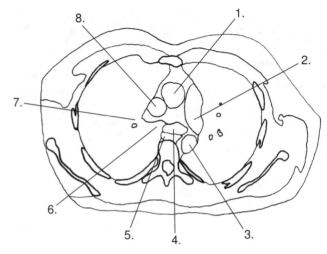

Figure 1-20

1. Which of the following is illustrated by 7?

    _____ A. Right main bronchus

    _____ B. Right lower lobe bronchus

    _____ C. Right upper lobe bronchus

    _____ D. Carina

2. What number illustrates the left main bronchus?

    _____ A. 7

    _____ B. 4

    _____ C. 5

    _____ D. 6

3. Which of the following is illustrated by 5?

    _____ A. Descending aorta

    _____ B. Carina

    _____ C. Left pulmonary artery

    _____ D. Ascending aorta

4. Which of the following is illustrated by 6?

    _____ A. Right main bronchus

    _____ B. Left main bronchus

    _____ C. Right lower lobe bronchus

    _____ D. Right upper lobe bronchus

5. What number illustrates the left pulmonary artery?

    _____ A. 8

    _____ B. 2

    _____ C. 3

    _____ D. 1

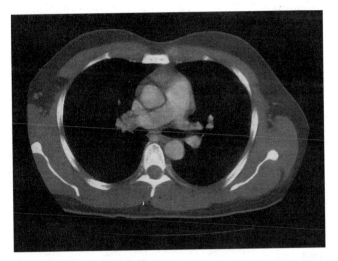

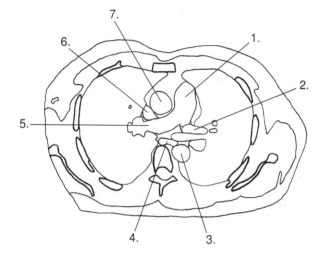

Figure 1-21

1. What number illustrates the esophagus?

_____ A. 7

_____ B. 4

_____ C. 3

_____ D. 2

2. Which of the following is illustrated by 5?

_____ A. Pulmonary trunk

_____ B. Left superior pulmonary vein

_____ C. Ascending aorta

_____ D. Right pulmonary artery

3. What number illustrates the left superior pulmonary vein?

_____ A. 1

_____ B. 2

_____ C. 5

_____ D. 3

4. Which of the following is illustrated by 1?

_____ A. Left pulmonary artery

_____ B. Ascending aorta

_____ C. Pulmonary trunk

_____ D. Descending aorta

5. Which of the following is illustrated by 6?

_____ A. Ascending aorta

_____ B. Left superior pulmonary vein

_____ C. Descending aorta

_____ D. Superior vena cava

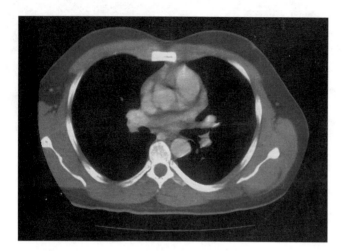
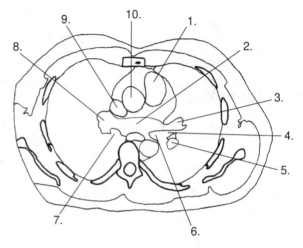

Figure 1-22

1. What number illustrates the left atrium?

    _____ A. 2

    _____ B. 9

    _____ C. 1

    _____ D. 10

2. Which of the following is illustrated by 7?

    _____ A. Left main bronchus

    _____ B. Bronchus intermedius

    _____ C. Superior pulmonary vein

    _____ D. Left upper lobe bronchus

3. What number illustrates the descending branch of the left pulmonary artery?

    _____ A. 6

    _____ B. 4

    _____ C. 5

    _____ D. 3

4. Which of the following is illustrated by 1?

    _____ A. Left superior pulmonary vein

    _____ B. Superior vena cava

    _____ C. Origin of the pulmonary trunk

    _____ D. Ascending aorta

5. Which of the following is illustrated by 4?

    _____ A. Left main bronchus

    _____ B. Descending branch of the left pulmonary artery

    _____ C. Left superior pulmonary vein

    _____ D. Left upper lobe bronchus

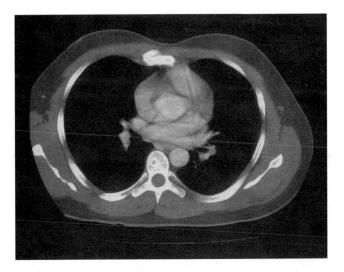 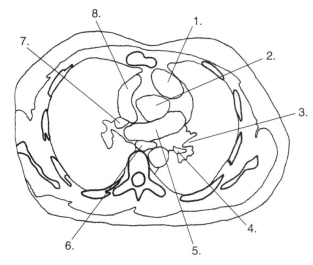

Figure 1-23

1. What number illustrates the ascending aorta?

   _____ A. 8

   _____ B. 2

   _____ C. 7

   _____ D. 1

2. Which of the following is illustrated by 4?

   _____ A. Left lower lobe bronchus

   _____ B. Left upper lobe bronchus

   _____ C. Left main bronchus

   _____ D. Carina

3. Which of the following is illustrated by 8?

   _____ A. Right ventricle

   _____ B. Right atrium

   _____ C. Right superior pulmonary vein

   _____ D. Left atrium

4. Which of the following is illustrated by 1?

   _____ A. Left atrium

   _____ B. Left ventricle

   _____ C. Right atrium

   _____ D. Right ventricle

5. What number illustrates the left atrium?

   _____ A. 8

   _____ B. 6

   _____ C. 5

   _____ D. 2

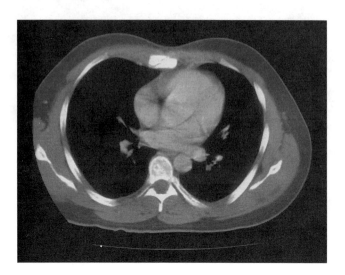
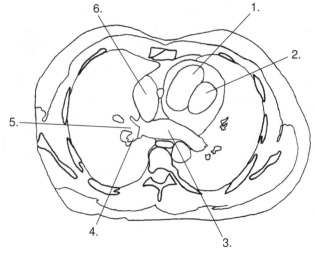

Figure 1-24

1. What number illustrates the left atrium?

    _____ A. 6

    _____ B. 2

    _____ C. 3

    _____ D. 1

2. Which of the following is illustrated by 4?

    _____ A. Right lower lobe bronchus

    _____ B. Right upper lobe bronchus

    _____ C. Right middle lobe of bronchus

    _____ D. Carina

3. Which of the following is illustrated by 5?

    _____ A. Right upper lobe bronchus

    _____ B. Right lower lobe bronchus

    _____ C. Right middle lobe bronchus

    _____ D. Carina

4. What number illustrates the chamber within the heart responsible for pumping blood to the lungs?

    _____ A. 1

    _____ B. 6

    _____ C. 2

    _____ D. 3

5. Which of the following is illustrated by 6?

    _____ A. Left atrium

    _____ B. Right ventricle

    _____ C. Left ventricle

    _____ D. Right atrium

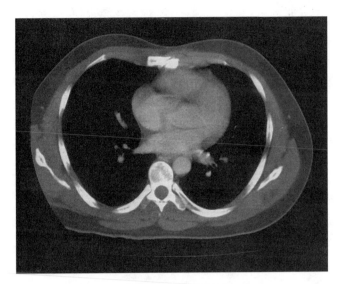 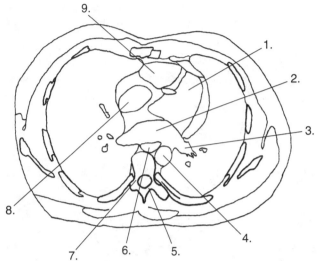

Figure 1-25

1. What number illustrates the lamina?

    _____ A. 7

    _____ B. 5

    _____ C. 6

    _____ D. 3

2. Which of the following is illustrated by 3?

    _____ A. Left atrium

    _____ B. Left inferior pulmonary vein

    _____ C. Left upper pulmonary vein

    _____ D. Hemiazygos vein

3. What number illustrates the right ventricle?

    _____ A. 8

    _____ B. 4

    _____ C. 1

    _____ D. 9

4. Which of the following is illustrated by 7?

    _____ A. Esophagus

    _____ B. Lamina

    _____ C. Pedicle

    _____ D. Descending aorta

5. Which of the following is illustrated by 1?

    _____ A. Left atrium

    _____ B. Right atrium

    _____ C. Left ventricle

    _____ D. Right ventricle

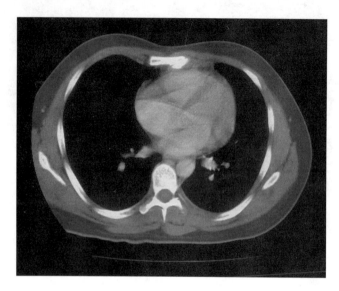 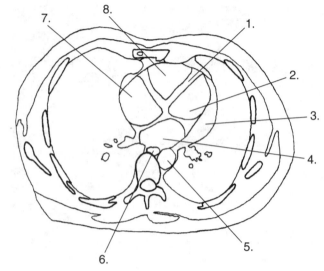

Figure 1-26

1. Which of the following is illustrated by 7?

    _____ A. Left atrium

    _____ B. Right atrium

    _____ C. Left ventricle

    _____ D. Right ventricle

2. Which of the following is illustrated by 8?

    _____ A. Right atrium

    _____ B. Left ventricle

    _____ C. Left atrium

    _____ D. Right ventricle

3. What number illustrates the descending aorta?

    _____ A. 4

    _____ B. 2

    _____ C. 5

    _____ D. 6

4. What number illustrates the ventricular wall?

    _____ A. 4

    _____ B. 3

    _____ C. 1

    _____ D. 6

5. Which of the following is illustrated by 4?

    _____ A. Left ventricle

    _____ B. Left atrium

    _____ C. Esophagus

    _____ D. Descending aorta

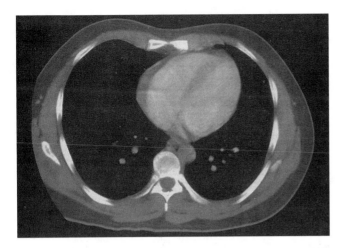 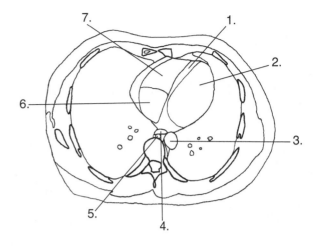

Figure 1-27

1. Which of the following is illustrated by 2?

_____ A. Interventricular septum

_____ B. Esophagus

_____ C. Descending aorta

_____ D. Left ventricle

2. What number illustrates the chamber of the heart receiving venous blood from the inferior vena cava?

_____ A. 3

_____ B. 2

_____ C. 6

_____ D. 7

3. Which of the following is illustrated by 5?

_____ A. Accessory hemiazygos vein

_____ B. Esophagus

_____ C. Azygos vein

_____ D. Hemiazygos vein

4. Which of the following is illustrated by 3?

_____ A. Descending aorta

_____ B. Hemiazygos vein

_____ C. Azygos vein

_____ D. Accessory hemiazygos vein

5. What number illustrates the right ventricle?

_____ A. 2

_____ B. 6

_____ C. 3

_____ D. 7

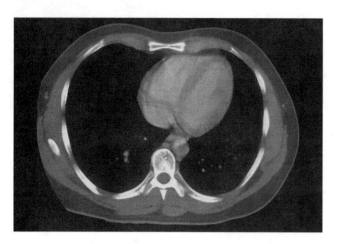
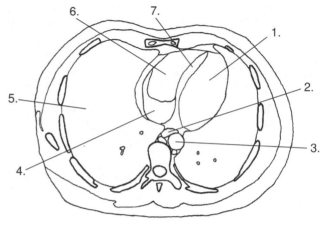

Figure 1-28

1. Which of the following is illustrated by 1?

   _____ A. Left atrium

   _____ B. Right ventricle

   _____ C. Left ventricle

   _____ D. Right atrium

2. Which of the following is illustrated by 5?

   _____ A. Left lung

   _____ B. Right lung

   _____ C. Right atrium

   _____ D. Right ventricle

3. What number illustrates the alimentary structure connecting the pharynx to the stomach?

   _____ A. 2

   _____ B. 3

   _____ C. 6

   _____ D. 4

4. What number illustrates the interventricular septum?

   _____ A. 2

   _____ B. 5

   _____ C. 4

   _____ D. 7

5. Which of the following is illustrated by 4?

   _____ A. Right atrium

   _____ B. Right ventricle

   _____ C. Left atrium

   _____ D. Left ventricle

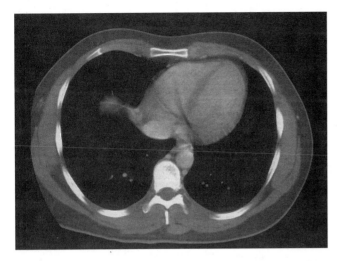 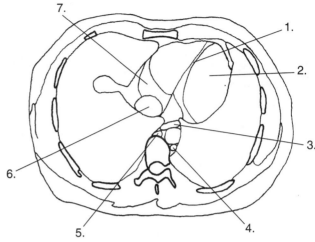

Figure 1-29

1. What number illustrates the inferior vena cava?

    _____ A. 3

    _____ B. 5

    _____ C. 4

    _____ D. 6

2. Which of the following is illustrated by 7?

    _____ A. Right ventricle

    _____ B. Right atrium

    _____ C. Inferior vena cava

    _____ D. Left ventricle

3. Which of the following is illustrated by 4?

    _____ A. Hemiazygos vein

    _____ B. Inferior vena cava

    _____ C. Azygos vein

    _____ D. Accessory hemiazygos vein

4. Which of the following is illustrated by 1?

    _____ A. Visceral peritoneum

    _____ B. Mediastinal pleura

    _____ C. Parietal peritoneum

    _____ D. Interventricular septum

5. What number illustrates the azygos vein?

    _____ A. 4

    _____ B. 3

    _____ C. 1

    _____ D. 5

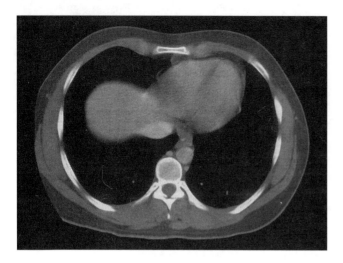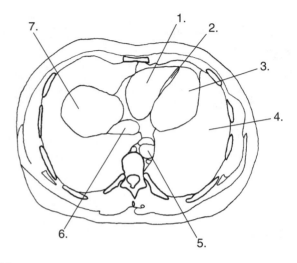

Figure 1-30

1. Which of the following is illustrated by 6?

    _____ A. Superior vena cava

    _____ B. Inferior vena cava

    _____ C. Right ventricle

    _____ D. Descending aorta

2. What number illustrates the interventricular septum?

    _____ A. 2

    _____ B. 6

    _____ C. 5

    _____ D. 4

3. What number illustrates the descending aorta?

    _____ A. 7

    _____ B. 3

    _____ C. 5

    _____ D. 6

4. Which of the following is illustrated by 7?

    _____ A. Left ventricle

    _____ B. Right ventricle

    _____ C. Stomach

    _____ D. Liver

5. Which of the following is illustrated by 3?

    _____ A. Left atrium

    _____ B. Left ventricle

    _____ C. Right ventricle

    _____ D. Liver

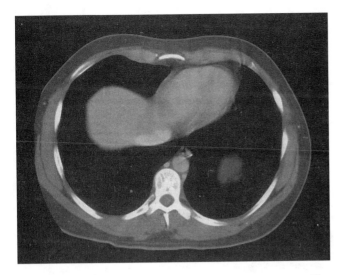
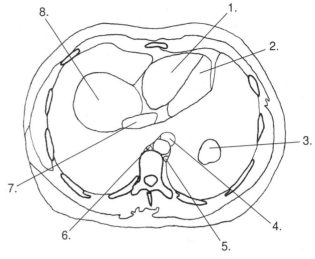

Figure 1-31

1. Which of the following is illustrated by 3?

   _____ A. Right ventricle

   _____ B. Left ventricle

   _____ C. Fundus of the stomach

   _____ D. Liver

2. What number illustrates the inferior vena cava?

   _____ A. 4

   _____ B. 6

   _____ C. 5

   _____ D. 7

3. Which of the following is illustrated by 6?

   _____ A. Azygos vein

   _____ B. Inferior vena cava

   _____ C. Hemiazygos vein

   _____ D. Accessory hemiazygos vein

4. What number illustrates the esophagus?

   _____ A. 3

   _____ B. 7

   _____ C. 6

   _____ D. 4

5. Which of the following is illustrated by 5?

   _____ A. Inferior vena cava

   _____ B. Accessory hemiazygos vein

   _____ C. Hemiazygos vein

   _____ D. Azygos vein

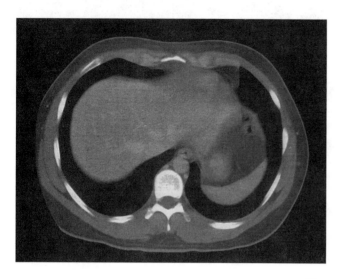 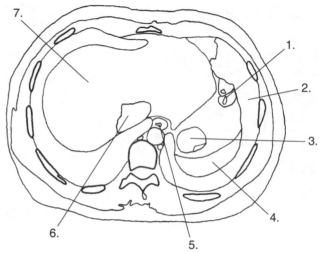

Figure 1-32

1. Which of the following is illustrated by 4?

    _____ A. Left lung

    _____ B. Spleen

    _____ C. Liver

    _____ D. Fundus of stomach

2. What number illustrates the vein transporting blood from the liver to the heart?

    _____ A. 5

    _____ B. 1

    _____ C. 3

    _____ D. 6

3. Which of the following is illustrated by 1?

    _____ A. Splenic flexure of the colon

    _____ B. Fundus of the stomach

    _____ C. Transverse colon

    _____ D. Spleen

4. Which of the following is illustrated by 3?

    _____ A. Spleen

    _____ B. Inferior vena cava

    _____ C. Esophagus

    _____ D. Fundus of the stomach

5. What number illustrates the left lung?

    _____ A. 2

    _____ B. 7

    _____ C. 6

    _____ D. 3

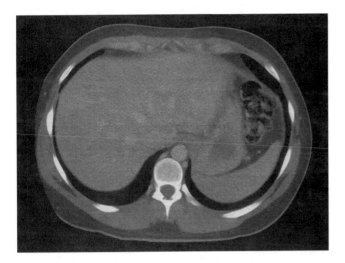
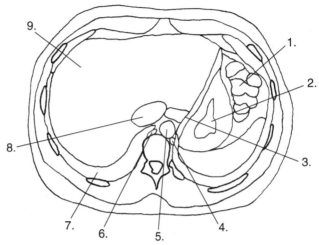

Figure 1-33

1. Which of the following is illustrated by 3?

　　_____ A. Gastroesophageal junction

　　_____ B. Spleen

　　_____ C. Splenic flexure of colon

　　_____ D. Liver

2. What number illustrates the splenic flexure of the colon?

　　_____ A. 1

　　_____ B. 8

　　_____ C. 6

　　_____ D. 3

3. Which of the following is illustrated by 6?

　　_____ A. Accessory hemiazygos vein

　　_____ B. Azygos vein

　　_____ C. Hemiazygos vein

　　_____ D. Inferior vena cava

4. Which of the following is illustrated by 5?

　　_____ A. Esophagus

　　_____ B. Inferior vena cava

　　_____ C. Descending aorta

　　_____ D. Azygos vein

5. What number illustrates the hemiazygos vein?

　　_____ A. 5

　　_____ B. 4

　　_____ C. 8

　　_____ D. 6

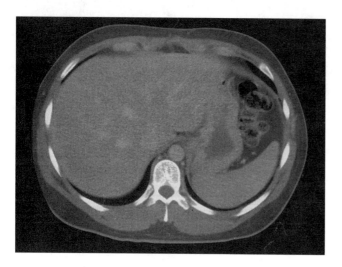 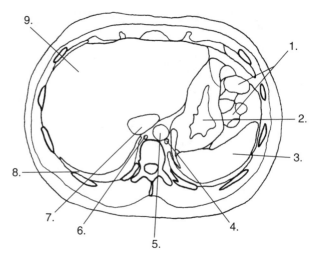

Figure 1-34

1. Which of the following is illustrated by 2?

_____ A. Splenic flexure of colon

_____ B. Fundus of stomach

_____ C. Spleen

_____ D. Body of stomach

2. What number illustrates the descending aorta?

_____ A. 4

_____ B. 2

_____ C. 5

_____ D. 6

3. Which of the following is illustrated by 3?

_____ A. Spleen

_____ B. Body of stomach

_____ C. Liver

_____ D. Splenic flexure of colon

4. Which of the following is illustrated by 7?

_____ A. Spleen

_____ B. Gallbladder

_____ C. Descending aorta

_____ D. Inferior vena cava

5. What number illustrates the splenic flexure of the colon?

_____ A. 1

_____ B. 5

_____ C. 3

_____ D. 2

ANSWERS

Fig. 1-1

1. B
2. C
3. D
4. B
5. A

Fig. 1-2

1. B
2. D
3. B
4. C
5. B

Fig. 1-3

1. C
2. C
3. C
4. B
5. C

Fig. 1-4

1. B
2. C
3. A
4. C
5. A

Fig. 1-5

1. C
2. D
3. C
4. A
5. B

Fig. 1-6

1. C
2. A
3. C
4. B
5. C

Fig. 1-7

1. C
2. A
3. C
4. C
5. D

Fig. 1-8

1. B
2. C
3. D
4. A
5. C

Fig. 1-9

1. D
2. B
3. A
4. B
5. C

Fig. 1-10

1. A
2. B
3. C
4. C
5. D

Fig. 1-11

1. C
2. B
3. B
4. D
5. A

Fig. 1-12

1. B
2. A
3. C
4. A
5. D

Fig. 1-13

1. B
2. B
3. A
4. C
5. C

Fig. 1-14

1. D
2. C
3. B
4. A
5. D

Fig. 1-15

1. B
2. D
3. D
4. C
5. C

Fig. 1-16

1. A
2. B
3. C
4. D
5. C

Fig. 1-17

1. D
2. C
3. A
4. B
5. B

Fig. 1-18

1. A
2. B
3. C
4. C
5. D

Fig. 1-19

1. B
2. C
3. D
4. A
5. B

Fig. 1-20

1. C
2. B
3. B
4. A
5. B

Fig. 1-21

1. B
2. D
3. B
4. C
5. D

Fig. 1-22

1. A
2. B
3. C
4. C
5. D

Fig. 1-23

1. B
2. A
3. B
4. D
5. C

Fig. 1-24

1. C
2. A
3. C
4. A
5. D

Fig. 1-25

1. B
2. B
3. D
4. C
5. C

Fig. 1-26

1. B
2. D
3. C
4. B
5. B

Fig. 1-27

1. D
2. C
3. C
4. A
5. D

Fig. 1-28

1. C
2. B
3. A
4. D
5. A

Fig. 1-29

1. D
2. B
3. A
4. D
5. D

Fig. 1-30

1. B
2. A
3. C
4. D
5. B

Fig. 1-31

1. C
2. D
3. A
4. D
5. C

Fig. 1-32

1. B
2. D
3. A
4. D
5. A

Fig. 1-33

1. A
2. A
3. B
4. C
5. B

Fig. 1-34

1. D
2. C
3. A
4. D
5. A

Chapter 2

Abdomen

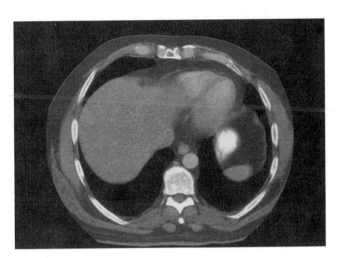

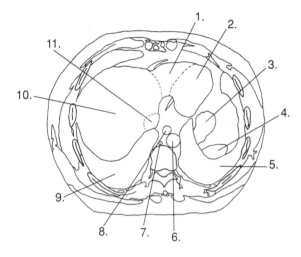

Figure 2-1

1. Which of the following is illustrated by 3?

_____ A. Left ventricle

_____ B. Inferior vena cava

_____ C. Fundus of stomach

_____ D. Spleen

2. Which of the following is illustrated by 10?

_____ A. Left ventricle

_____ B. Spleen

_____ C. Right lung

_____ D. Liver

3. What number illustrates the azygos vein?

_____ A. 11

_____ B. 6

_____ C. 7

_____ D. 3

4. What number illustrates the inferior vena cava?

_____ A. 6

_____ B. 8

_____ C. 3

_____ D. 11

5. Which of the following is illustrated by 4?

_____ A. Spleen

_____ B. Descending aorta

_____ C. Esophagus

_____ D. Inferior vena cava

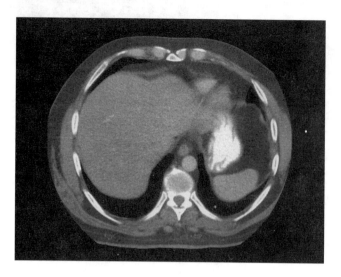 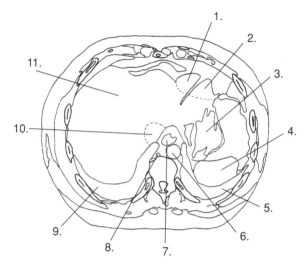

Figure 2-2

1. What number illustrates the chamber of the heart that pumps blood to most of the body?

    _____ A. 6

    _____ B. 2

    _____ C. 3

    _____ D. 1

2. Which of the following is illustrated by 5?

    _____ A. Right lung

    _____ B. Spleen

    _____ C. Left lung

    _____ D. Inferior vena cava

3. Which of the following is illustrated by 7?

    _____ A. Inferior vena cava

    _____ B. Esophagus

    _____ C. Descending aorta

    _____ D. Azygos vein

4. What number illustrates the liver?

    _____ A. 9

    _____ B. 4

    _____ C. 3

    _____ D. 11

5. Which of the following is illustrated by 1?

    _____ A. Quadrate lobe of liver

    _____ B. Right ventricle

    _____ C. Left ventricle

    _____ D. Gallbladder

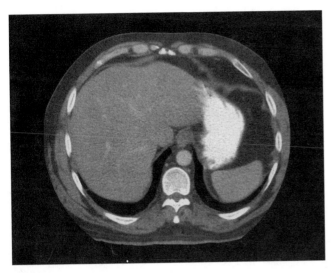
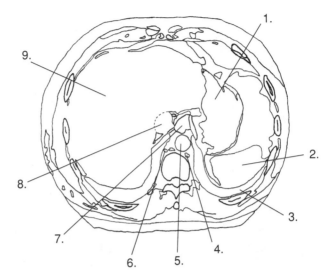

Figure 2-3

1. Which of the following is illustrated by 6?

_____ A. Hemiazygos vein

_____ B. Portal vein

_____ C. Accessory hemiazygos vein

_____ D. Azygos vein

2. Which of the following is illustrated by 1?

_____ A. Spleen

_____ B. Fundus of stomach

_____ C. Quadrate lobe of liver

_____ D. Esophagus

3. What number illustrates the hemiazygos vein?

_____ A. 4

_____ B. 7

_____ C. 6

_____ D. 5

4. Which of the following is illustrated by 8?

_____ A. Descending aorta

_____ B. Fundus of stomach

_____ C. Inferior vena cava

_____ D. Esophagus

5. What number illustrates the left lung?

_____ A. 9

_____ B. 2

_____ C. 3

_____ D. 1

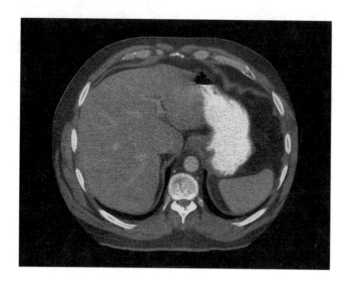
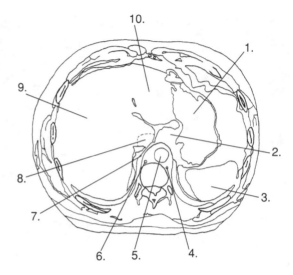

Figure 2-4

1. What number illustrates the body of the stomach?

_____ A. 1

_____ B. 9

_____ C. 5

_____ D. 2

2. Which of the following is illustrated by 7?

_____ A. Hemiazygos vein

_____ B. Right crus of diaphragm

_____ C. Azygos vein

_____ D. Right adrenal gland

3. What number illustrates the right lobe of the liver?

_____ A. 10

_____ B. 3

_____ C. 9

_____ D. 1

4. Which of the following is illustrated by 2?

_____ A. Pancreas

_____ B. Portal vein

_____ C. Inferior vena cava

_____ D. Gastroesophageal junction

5. Which of the following is illustrated by 6?

_____ A. Hemiazygos vein

_____ B. Azygos vein

_____ C. Portal vein

_____ D. Right crus of diaphragm

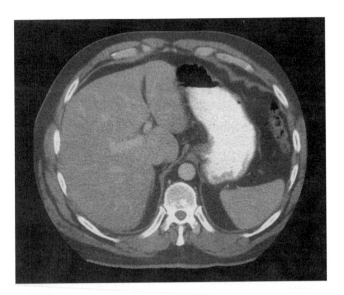 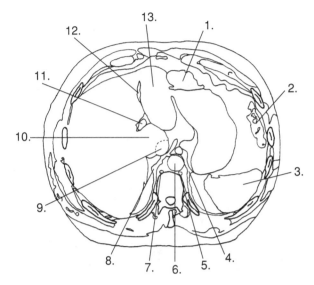

Figure 2-5

1. Which of the following is illustrated by 12?

    _____ A. Quadrate lobe of liver

    _____ B. Caudate lobe of liver

    _____ C. Inferior vena cava

    _____ D. Ligamentum teres fossa

2. What number illustrates the caudate lobe of liver?

    _____ A. 13

    _____ B. 3

    _____ C. 10

    _____ D. 1

3. Which of the following is illustrated by 2?

    _____ A. Air in stomach

    _____ B. Transverse colon

    _____ C. Ascending colon

    _____ D. Splenic flexure of colon

4. Which of the following is illustrated by 6?

    _____ A. Descending aorta

    _____ B. Celiac trunk

    _____ C. Inferior vena cava

    _____ D. Portal vein

5. What number illustrates the left lobe of liver?

    _____ A. 1

    _____ B. 10

    _____ C. 6

    _____ D. 13

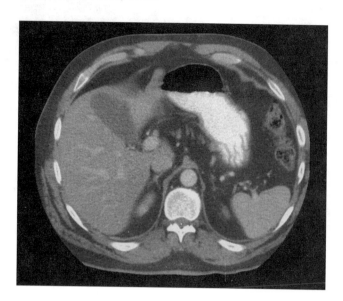
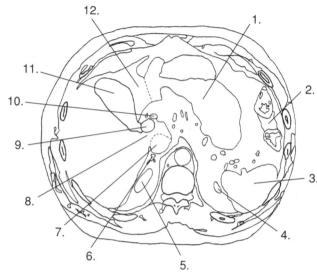

Figure 2-6

1. Which of the following is illustrated by 4?

    _____ A. Splenic vein

    _____ B. Upper pole of left kidney

    _____ C. Left adrenal gland

    _____ D. Inferior mesenteric vein

2. What number illustrates the adrenal gland?

    _____ A. 6

    _____ B. 9

    _____ C. 7

    _____ D. 5

3. What number illustrates the gallbladder?

    _____ A. 7

    _____ B. 9

    _____ C. 12

    _____ D. 11

4. Which of the following is illustrated by 12?

    _____ A. Left lobe of liver

    _____ B. Caudate lobe of liver

    _____ C. Gallbladder

    _____ D. Quadrate lobe of liver

5. Which of the following is illustrated by 10?

    _____ A. Common hepatic artery

    _____ B. Proper hepatic artery

    _____ C. Portal vein

    _____ D. Gastroduodenal artery

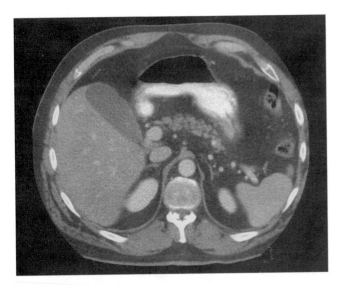 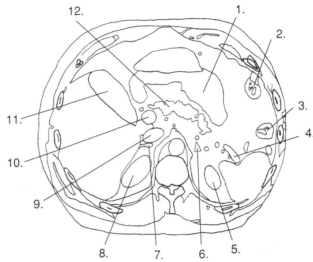

Figure 2-7

1. What number illustrates the descending colon?

    _____ A. 9

    _____ B. 3

    _____ C. 1

    _____ D. 2

2. Which of the following is illustrated by 7?

    _____ A. Upper pole of right kidney

    _____ B. Inferior vena cava

    _____ C. Splenic vein

    _____ D. Right adrenal gland

3. Which of the following is illustrated by 10?

    _____ A. Transverse colon

    _____ B. Inferior vena cava

    _____ C. Portal vein

    _____ D. Proper hepatic artery

4. Which of the following is illustrated by 2?

    _____ A. Descending colon

    _____ B. Splenic vein

    _____ C. Transverse colon

    _____ D. Left adrenal

5. What number illustrates the body of the pancreas?

    _____ A. 11

    _____ B. 1

    _____ C. 8

    _____ D. 12

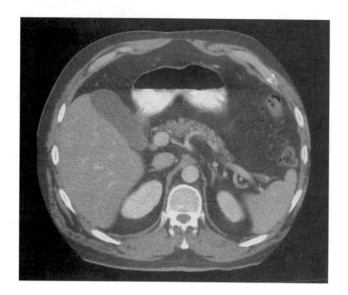 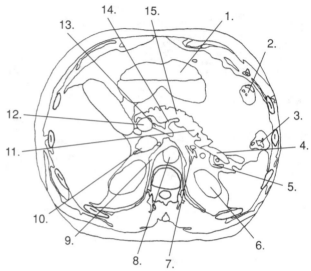

Figure 2-8

1. Which of the following is illustrated by 13?

    _____ A. Hepatic artery

    _____ B. Splenic artery

    _____ C. Celiac trunk

    _____ D. Cystic duct

2. What number illustrates the transverse colon?

    _____ A. 2

    _____ B. 3

    _____ C. 1

    _____ D. 15

3. Which of the following is illustrated by 11?

    _____ A. Body of pancreas

    _____ B. Stomach

    _____ C. Duodenum

    _____ D. Tail of pancreas

4. Which of the following is illustrated by 6?

    _____ A. Left adrenal gland

    _____ B. Left kidney

    _____ C. Splenic flexure of colon

    _____ D. Spleen

5. What number illustrates the splenic artery?

    _____ A. 13

    _____ B. 5

    _____ C. 4

    _____ D. 7

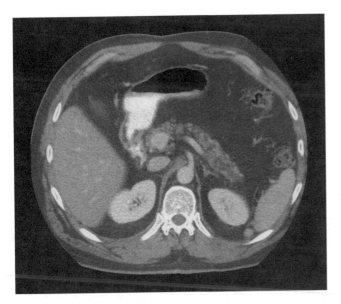
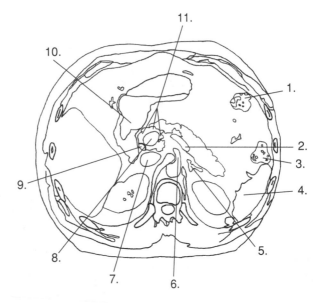

Figure 2-9

1. Which of the following is illustrated by 9?

    _____ A. Superior mesenteric artery

    _____ B. Hepatic artery

    _____ C. Portal vein

    _____ D. Common bile duct

2. What number illustrates the celiac trunk?

    _____ A. 8

    _____ B. 6

    _____ C. 5

    _____ D. 7

3. Which of the following is illustrated by 4?

    _____ A. Left adrenal gland

    _____ B. Inferior vena cava

    _____ C. Spleen

    _____ D. Head of pancreas

4. What number illustrates the pyloric antrum of the stomach?

    _____ A. 6

    _____ B. 1

    _____ C. 3

    _____ D. 10

5. Which of the following is illustrated by 11?

    _____ A. Head of pancreas

    _____ B. Pyloric antrum of stomach

    _____ C. Common bile duct

    _____ D. Duodenum

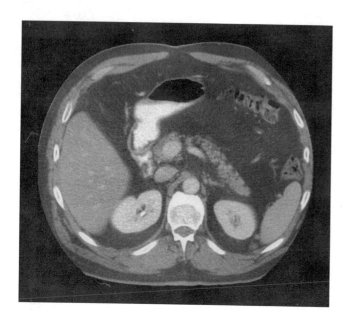 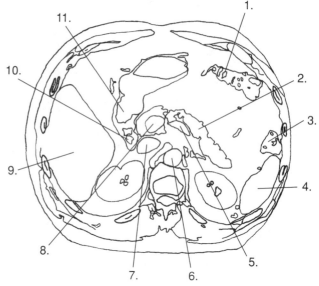

Figure 2-10

1. Which of the following is illustrated by 2?

    _____ A. Splenic vein

    _____ B. Body of pancreas

    _____ C. Portal vein

    _____ D. Duodenum

2. What number illustrates the first part of the small bowel?

    _____ A. 7

    _____ B. 2

    _____ C. 11

    _____ D. 10

3. What number illustrates the descending or abdominal aorta?

    _____ A. 7

    _____ B. 2

    _____ C. 6

    _____ D. 5

4. Which of the following is illustrated by 9?

    _____ A. Liver

    _____ B. Spleen

    _____ C. Body of pancreas

    _____ D. Pyloric antrum of stomach

5. Which of the following is illustrated by 1?

    _____ A. Descending colon

    _____ B. Transverse colon

    _____ C. Ascending colon

    _____ D. Duodenum

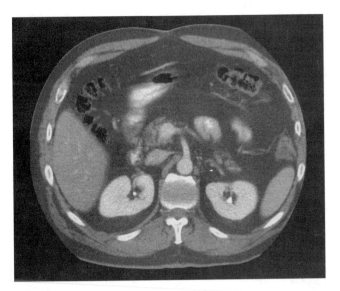
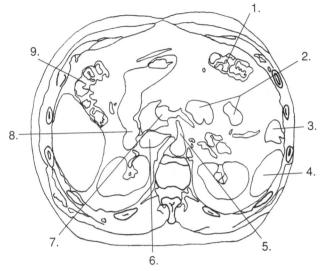

Figure 2-11

1. What number illustrates the vein carrying nutrient-rich venous blood to the liver?

    _____ A. 6

    _____ B. 5

    _____ C. 8

    _____ D. 7

2. Which of the following is illustrated by 9?

    _____ A. Transverse colon

    _____ B. Duodenum

    _____ C. Hepatic flexure of colon

    _____ D. Descending colon

3. Which of the following is illustrated by 5?

    _____ A. Celiac trunk

    _____ B. Superior mesenteric artery

    _____ C. Inferior mesenteric artery

    _____ D. Inferior vena cava

4. What number illustrates the descending colon?

    _____ A. 3

    _____ B. 1

    _____ C. 9

    _____ D. 4

5. Which of the following is illustrated by 4?

    _____ A. Adrenal gland

    _____ B. Ascending colon

    _____ C. Spleen

    _____ D. Liver

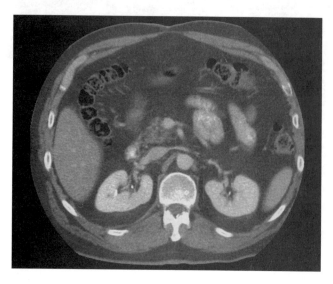 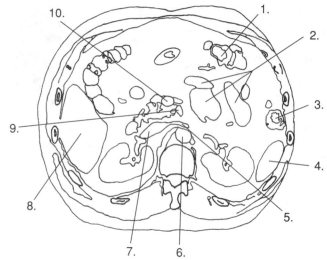

Figure 2-12

1. Which of the following is illustrated by 10?
   _____ A. Portal vein
   _____ B. Common bile duct
   _____ C. Superior mesenteric artery
   _____ D. Superior mesenteric vein

2. Which of the following is illustrated by 5?
   _____ A. Inferior mesenteric vein
   _____ B. Left renal vein
   _____ C. Crus of diaphragm
   _____ D. Superior mesenteric vein

3. What number illustrates the loops of the small bowel?
   _____ A. 3
   _____ B. 2
   _____ C. 4
   _____ D. 1

4. What number illustrates the superior mesenteric artery?
   _____ A. 10
   _____ B. 5
   _____ C. 9
   _____ D. 7

5. Which of the following is illustrated by 3?
   _____ A. Loops of small bowel
   _____ B. Ascending colon
   _____ C. Transverse colon
   _____ D. Descending colon

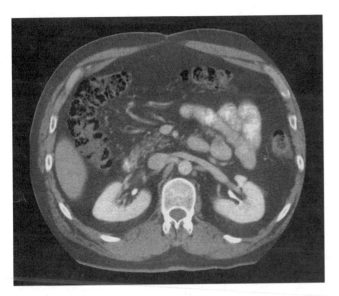 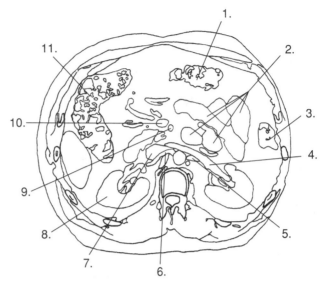

Figure 2-13

1. Which of the following is illustrated by 4?

    _____ A. Left renal vein

    _____ B. Left renal artery

    _____ C. Inferior mesenteric vein

    _____ D. Inferior mesenteric artery

2. What number illustrates the inferior vena cava?

    _____ A. 6

    _____ B. 9

    _____ C. 7

    _____ D. 5

3. Which of the following is illustrated by 8?

    _____ A. Adrenal gland

    _____ B. Kidney

    _____ C. Loops of small bowel

    _____ D. Hepatic flexure of colon

4. What number illustrates the hepatic flexure of the colon?

    _____ A. 3

    _____ B. 8

    _____ C. 1

    _____ D. 11

5. Which of the following is illustrated by 5?

    _____ A. Renal vein

    _____ B. Renal artery

    _____ C. Superior mesenteric vein

    _____ D. Superior mesenteric artery

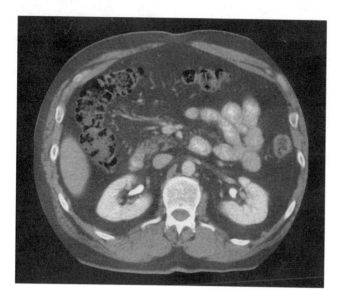
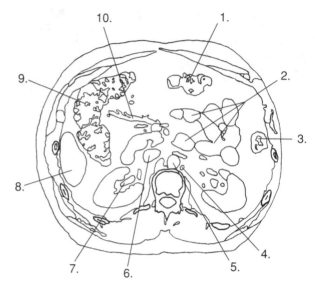

Figure 2-14

1. Which of the following is illustrated by 8?
   _____ A. Descending colon
   _____ B. Spleen
   _____ C. Liver
   _____ D. Adrenal gland

2. What number illustrates the hemiazygos vein?
   _____ A. 4
   _____ B. 5
   _____ C. 7
   _____ D. 6

3. What number illustrates the right renal pelvis?
   _____ A. 8
   _____ B. 6
   _____ C. 7
   _____ D. 5

4. Which of the following is illustrated by 10?
   _____ A. Portal vein
   _____ B. Loops of small bowel
   _____ C. Common bile duct
   _____ D. Mesenteric vessels

5. Which of the following is illustrated by 5?
   _____ A. Hemiazygos vein
   _____ B. Descending aorta
   _____ C. Inferior vena cava
   _____ D. Inferior mesenteric artery

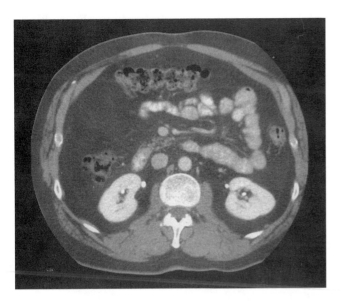 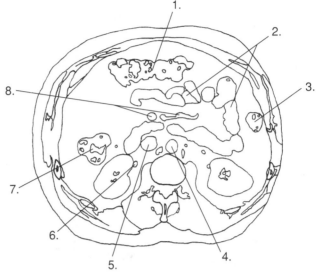

Figure 2-15

1. What number illustrates the abdominal aorta?

    _____ A. 4

    _____ B. 8

    _____ C. 5

    _____ D. 6

2. Which of the following is illustrated by 6?

    _____ A. Right renal vein

    _____ B. Right renal artery

    _____ C. Right ureter

    _____ D. Azygos vein

3. Which of the following is illustrated by 8?

    _____ A. Celiac vessels

    _____ B. Superior mesenteric vessels

    _____ C. Inferior mesenteric vessels

    _____ D. Loops of small bowel

4. Which of the following is illustrated by 3?

    _____ A. Ascending colon

    _____ B. Small bowel

    _____ C. Cecum

    _____ D. Descending colon

5. What number illustrates the ascending colon?

    _____ A. 2

    _____ B. 1

    _____ C. 3

    _____ D. 7

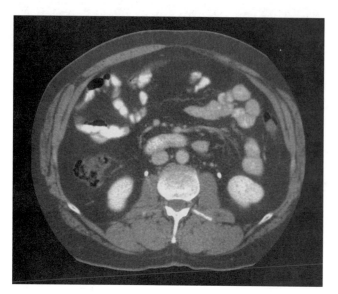 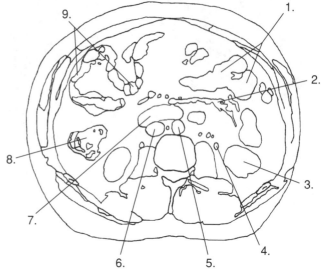

Figure 2-16

1. Which of the following is illustrated by 1?

    _____ A. Ascending colon

    _____ B. Transverse colon

    _____ C. Descending colon

    _____ D. Loops of small bowel

2. Which of the following is illustrated by 3?

    _____ A. Adrenal gland

    _____ B. Lower pole of left kidney

    _____ C. Ascending colon

    _____ D. Spleen

3. What number illustrates the duodenum?

    _____ A. 3

    _____ B. 1

    _____ C. 6

    _____ D. 7

4. What number illustrates the inferior vena cava?

    _____ A. 5

    _____ B. 4

    _____ C. 6

    _____ D. 7

5. Which of the following is illustrated by 4?

    _____ A. Left ureter

    _____ B. Left renal vein

    _____ C. Left renal artery

    _____ D. Hemiazygos vein

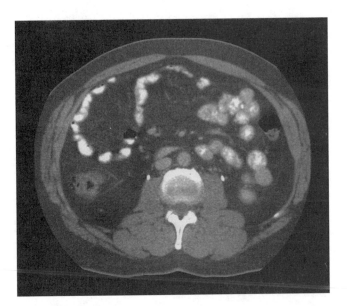 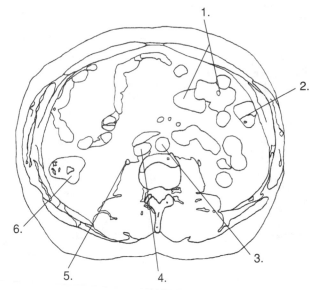

Figure 2-17

1. What number illustrates the abdominal aorta?

   _____ A. 3

   _____ B. 6

   _____ C. 5

   _____ D. 4

2. Which of the following is illustrated by 6?

   _____ A. Small bowel

   _____ B. Ascending colon

   _____ C. Descending colon

   _____ D. Cecum

3. What number illustrates the inferior vena cava?

   _____ A. 5

   _____ B. 2

   _____ C. 4

   _____ D. 3

4. Which of the following is illustrated by 2?

   _____ A. Loops of small bowel

   _____ B. Ascending colon

   _____ C. Cecum

   _____ D. Descending colon

5. Which of the following is illustrated by 5?

   _____ A. Inferior mesenteric artery

   _____ B. Right ureter

   _____ C. Inferior mesenteric vein

   _____ D. Right renal artery

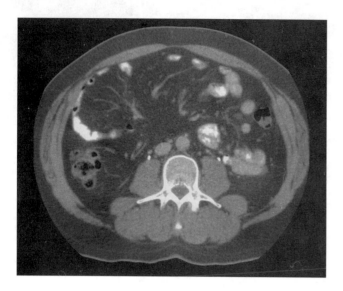
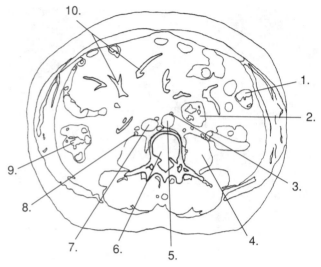

Figure 2-18

1. Which of the following is illustrated by 3?

    _____ A. Superior mesenteric artery

    _____ B. Inferior mesenteric artery

    _____ C. Celiac trunk

    _____ D. Ureter

2. Which of the following is illustrated by 1?

    _____ A. Ascending colon

    _____ B. Small bowel

    _____ C. Descending colon

    _____ D. Cecum

3. What number illustrates the ascending colon?

    _____ A. 10

    _____ B. 1

    _____ C. 2

    _____ D. 9

4. Which of the following is illustrated by 2?

    _____ A. Small bowel

    _____ B. Descending colon

    _____ C. Transverse colon

    _____ D. Ascending colon

5. What number illustrates the inferior mesenteric artery?

    _____ A. 7

    _____ B. 10

    _____ C. 8

    _____ D. 3

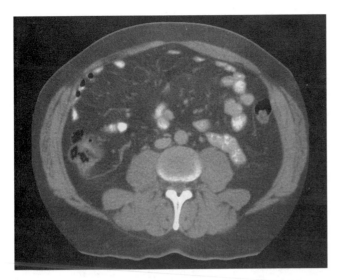
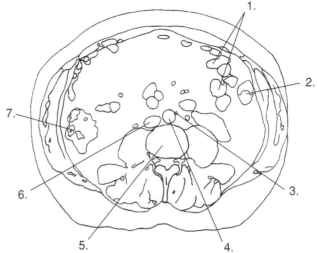

Figure 2-19

1. What number illustrates the loops of the small bowel?

    _____ A. 1

    _____ B. 7

    _____ C. 2

    _____ D. 4

2. Which of the following is illustrated by 5?

    _____ A. Psoas muscle

    _____ B. Vertebral body of L3

    _____ C. Intervertebral disk

    _____ D. Vertebral body of L4

3. Which of the following is illustrated by 7?

    _____ A. Descending colon

    _____ B. Ascending colon

    _____ C. Small bowel

    _____ D. Cecum

4. What number illustrates the abdominal aorta?

    _____ A. 2

    _____ B. 1

    _____ C. 4

    _____ D. 6

5. Which of the following is illustrated by 2?

    _____ A. Descending colon

    _____ B. Small bowel

    _____ C. Ascending colon

    _____ D. Cecum

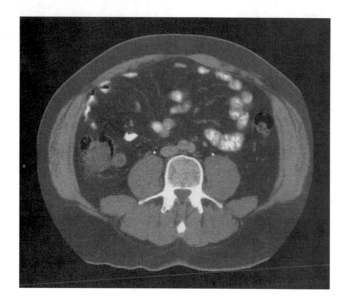 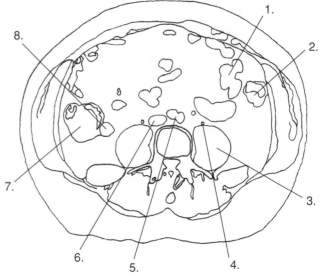

Figure 2-20

1. Which of the following is illustrated by 3?

_____ A. Iliacus muscle

_____ B. Psoas muscle

_____ C. Ascending colon

_____ D. Cecum

2. What number illustrates the common iliac arteries?

_____ A. 6

_____ B. 2

_____ C. 5

_____ D. 4

3. Which of the following is illustrated by 8?

_____ A. Cecum

_____ B. Descending colon

_____ C. Ileum

_____ D. Jejunum

4. Which of the following is illustrated by 7?

_____ A. Descending colon

_____ B. Ascending colon

_____ C. Small bowel

_____ D. Cecum

5. What number illustrates the small bowel?

_____ A. 1

_____ B. 3

_____ C. 2

_____ D. 7

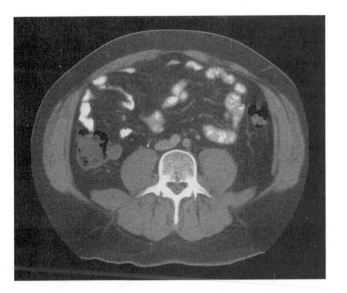 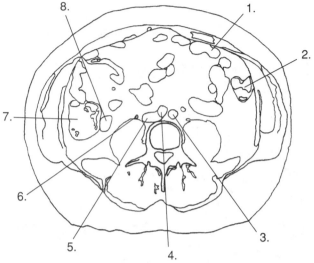

Figure 2-21

1. What number illustrates the ureter?

    _____ A. 4

    _____ B. 3

    _____ C. 5

    _____ D. 6

2. Which of the following is illustrated by 8?

    _____ A. Right common iliac artery

    _____ B. Cecum

    _____ C. Ileum

    _____ D. Jejunum

3. Which of the following is illustrated by 4?

    _____ A. Inferior mesenteric artery

    _____ B. Right common iliac artery

    _____ C. Inferior mesenteric vein

    _____ D. Inferior vena cava

4. Which of the following is illustrated by 2?

    _____ A. Small bowel

    _____ B. Descending colon

    _____ C. Ascending colon

    _____ D. Cecum

5. What number illustrates the left common iliac artery?

    _____ A. 3

    _____ B. 8

    _____ C. 4

    _____ D. 5

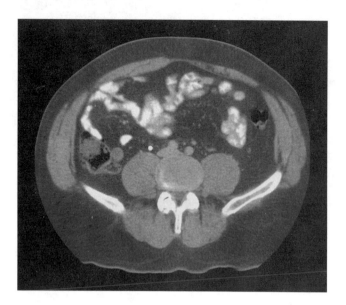
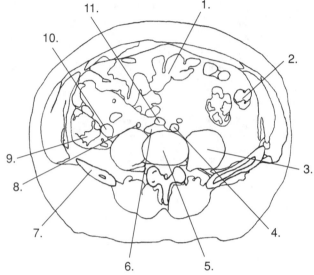

Figure 2-22

1. Which of the following is illustrated by 8?

    _____ A. Left common iliac artery

    _____ B. Ureter

    _____ C. Inferior mesenteric artery

    _____ D. Right common iliac artery

2. What number illustrates the psoas muscle?

    _____ A. 3

    _____ B. 10

    _____ C. 7

    _____ D. 5

3. What number illustrates the iliac crest?

    _____ A. 9

    _____ B. 1

    _____ C. 7

    _____ D. 4

4. Which of the following is illustrated by 5?

    _____ A. Vertebral body of L3

    _____ B. Intervertebral disk

    _____ C. Vertebral body of L4

    _____ D. Sacrum

5. Which of the following is illustrated by 6?

    _____ A. Left common iliac artery

    _____ B. Right common iliac artery

    _____ C. Inferior mesenteric vein

    _____ D. Inferior vena cava

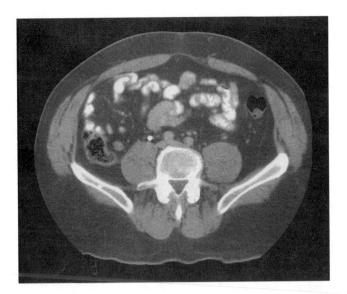 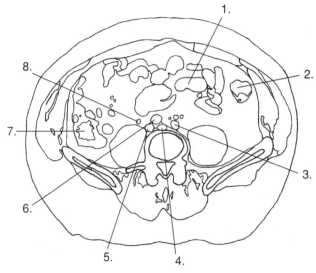

Figure 2-23

1. What number illustrates the small bowel?

    _____ A. 1

    _____ B. 6

    _____ C. 2

    _____ D. 7

2. Which of the following is illustrated by 4?

    _____ A. Left common iliac artery

    _____ B. Right common iliac vein

    _____ C. Left common iliac vein

    _____ D. Right common iliac artery

3. What number illustrates the right common iliac vein?

    _____ A. 6

    _____ B. 4

    _____ C. 5

    _____ D. 3

4. Which of the following is illustrated by 8?

    _____ A. Right common iliac vein

    _____ B. Right common iliac artery

    _____ C. Right ureter

    _____ D. Inferior mesenteric vein

5. Which of the following is illustrated by 2?

    _____ A. Cecum

    _____ B. Ascending colon

    _____ C. Loops of small bowel

    _____ D. Descending colon

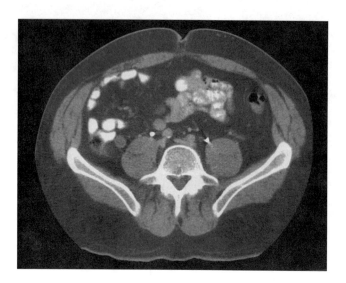 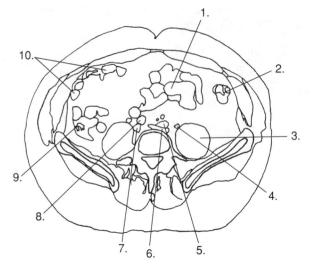

Figure 2-24

1. Which of the following is illustrated by 3?

    _____ A. Small bowel

    _____ B. Psoas muscle

    _____ C. Right common iliac vein

    _____ D. Cecum

2. What number illustrates the cecum?

    _____ A. 2

    _____ B. 3

    _____ C. 1

    _____ D. 9

3. What number illustrates the ureter?

    _____ A. 4

    _____ B. 8

    _____ C. 7

    _____ D. 5

4. Which of the following is illustrated by 1?

    _____ A. Cecum

    _____ B. Left psoas muscle

    _____ C. Small bowel

    _____ D. Transverse colon

5. Which of the following is illustrated by 7?

    _____ A. Ureter

    _____ B. Right common iliac vein

    _____ C. Right internal iliac vein

    _____ D. Right common iliac artery

ANSWERS

Fig. 2-1

1. C
2. D
3. C
4. D
5. A

Fig. 2-2

1. B
2. C
3. B
4. D
5. B

Fig. 2-3

1. D
2. B
3. A
4. C
5. C

Fig. 2-4

1. A
2. B
3. C
4. D
5. B

Fig. 2-5

1. D
2. C
3. D
4. A
5. D

Fig. 2-6

1. B
2. A
3. D
4. D
5. B

Fig. 2-7

1. B
2. D
3. C
4. C
5. D

Fig. 2-8

1. A
2. A
3. D
4. B
5. B

Fig. 2-9

1. D
2. B
3. C
4. D
5. A

Fig. 2-10

1. B
2. D
3. C
4. A
5. B

Fig. 2-11

1. D
2. C
3. B
4. A
5. C

Fig. 2-12

1. D
2. B
3. B
4. C
5. D

Fig. 2-13

1. A
2. C
3. B
4. D
5. B

Fig. 2-14

1. C
2. A
3. C
4. D
5. B

Fig. 2-15

1. A
2. C
3. B
4. D
5. D

Fig. 2-16

1. D
2. B
3. D
4. C
5. A

Fig. 2-17

1. A
2. B
3. C
4. D
5. B

Fig. 2-18

1. B
2. C
3. D
4. A
5. D

Fig. 2-19

1. A
2. C
3. B
4. C
5. A

Fig. 2-20

1. B
2. C
3. C
4. D
5. A

Fig. 2-21

1. D
2. C
3. B
4. B
5. A

Fig. 2-22

1. B
2. A
3. C
4. B
5. D

Fig. 2-23

1. A
2. C
3. C
4. C
5. D

Fig. 2-24

1. B
2. D
3. A
4. C
5. B

Chapter 3

Male and Female Pelvis

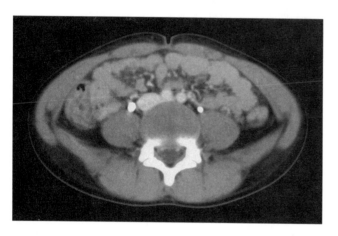
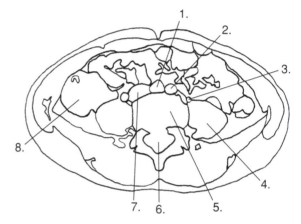

Figure 3-1

1. Which of the following is illustrated by 2?

    _____ A. Abdominal aorta

    _____ B. Inferior vena cava

    _____ C. Common iliac artery

    _____ D. Common iliac vein

2. Which of the following is illustrated by 6?

    _____ A. Intervertebral foramen

    _____ B. Vertebral foramen

    _____ C. Vertebral body

    _____ D. Intervertebral disk

3. What number illustrates the psoas muscle?

    _____ A. 7

    _____ B. 8

    _____ C. 4

    _____ D. 1

4. Which of the following is illustrated by 7?

    _____ A. Common iliac vein

    _____ B. Common iliac artery

    _____ C. Abdominal aorta

    _____ D. Inferior vena cava

5. What number illustrates the intervertebral disk?

    _____ A. 8

    _____ B. 5

    _____ C. 6

    _____ D. 4

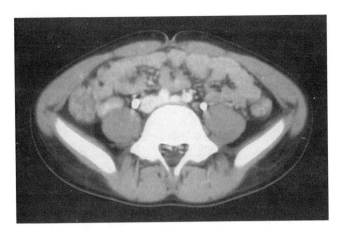 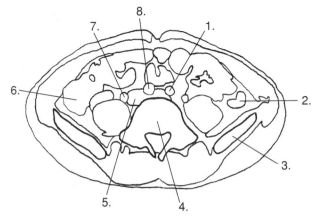

Figure 3-2

1. What number illustrates the iliac crest?

_____ A. 4

_____ B. 2

_____ C. 3

_____ D. 1

2. Which of the following is illustrated by 8?

_____ A. Inferior vena cava

_____ B. Common iliac artery

_____ C. Common iliac vein

_____ D. Abdominal aorta

3. Which of the following is illustrated by 7?

_____ A. Right ureter

_____ B. Common iliac artery

_____ C. Common iliac vein

_____ D. Inferior mesenteric artery

4. What number illustrates the inferior vena cava?

_____ A. 8

_____ B. 7

_____ C. 1

_____ D. 5

5. What number illustrates the descending colon?

_____ A. 5

_____ B. 2

_____ C. 7

_____ D. 6

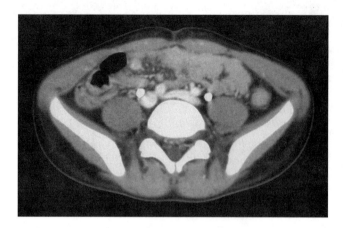
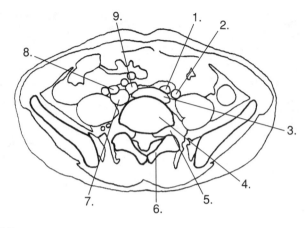

Figure 3-3

1. What number illustrates the right common iliac artery?
   _____ A. 7
   _____ B. 2
   _____ C. 1
   _____ D. 9

2. Which of the following is illustrated by 7?
   _____ A. Left internal iliac vein
   _____ B. Right common iliac vein
   _____ C. Left external iliac artery
   _____ D. Right common iliac artery

3. Which of the following is illustrated by 3?
   _____ A. Abdominal aorta
   _____ B. Left common iliac artery
   _____ C. Inferior vena cava
   _____ D. Left common iliac vein

4. What number illustrates the lamina?
   _____ A. 1
   _____ B. 5
   _____ C. 2
   _____ D. 6

5. Which of the following is illustrated by 8?
   _____ A. External iliac artery
   _____ B. Common iliac artery
   _____ C. Common iliac vein
   _____ D. Ureter

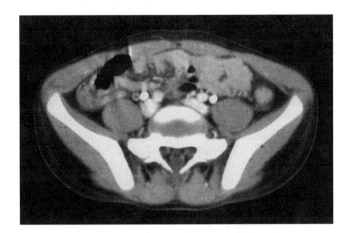 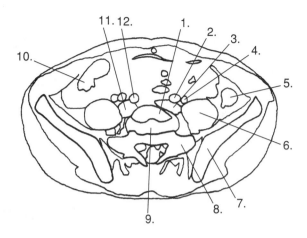

Figure 3-4

1. What number illustrates the ilium?

    _____ A. 6

    _____ B. 5

    _____ C. 10

    _____ D. 7

2. Which of the following is illustrated by 1?

    _____ A. S1

    _____ B. L5

    _____ C. Intervertebral disk

    _____ D. L4

3. What number illustrates the cecum?

    _____ A. 11

    _____ B. 7

    _____ C. 10

    _____ D. 5

4. Which of the following is illustrated by 2?

    _____ A. Left common iliac artery

    _____ B. Left common iliac vein

    _____ C. Left external iliac artery

    _____ D. Left external iliac vein

5. What number illustrates the psoas muscle?

    _____ A. 7

    _____ B. 6

    _____ C. 5

    _____ D. 11

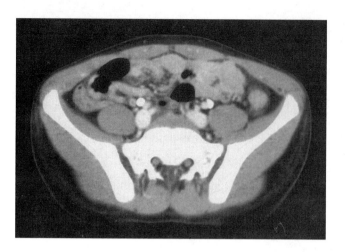
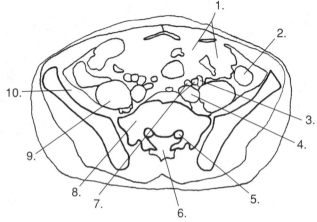

Figure 3-5

1. What number illustrates the left common iliac vein?

    _____ A. 5

    _____ B. 7

    _____ C. 4

    _____ D. 3

2. Which of the following is illustrated by 10?

    _____ A. Loops of small bowel

    _____ B. Right ilium

    _____ C. Right iliacus muscle

    _____ D. Right psoas muscle

3. Which of the following is illustrated by 5?

    _____ A. Left ureter

    _____ B. Left vertebral foramen

    _____ C. Left common iliac vein

    _____ D. Sacral canal

4. What number illustrates the left ureter?

    _____ A. 5

    _____ B. 3

    _____ C. 7

    _____ D. 2

5. What number illustrates the psoas muscle?

    _____ A. 10

    _____ B. 4

    _____ C. 9

    _____ D. 6

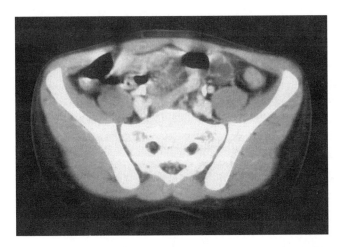
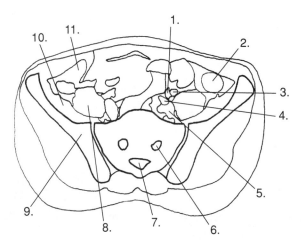

Figure 3-6

1. Which of the following is illustrated by 11?

    _____ A. Ileum

    _____ B. Cecum

    _____ C. Descending colon

    _____ D. Ascending colon

2. What number illustrates the left internal iliac artery?

    _____ A. 5

    _____ B. 3

    _____ C. 4

    _____ D. 1

3. What number illustrates the left external iliac artery?

    _____ A. 3

    _____ B. 4

    _____ C. 1

    _____ D. 5

4. Which of the following is illustrated by 2?

    _____ A. Jejunum

    _____ B. Ileum

    _____ C. Cecum

    _____ D. Descending colon

5. Which of the following is illustrated by 3?

    _____ A. Left common iliac vein

    _____ B. Left external iliac artery

    _____ C. Left internal iliac artery

    _____ D. Left ureter

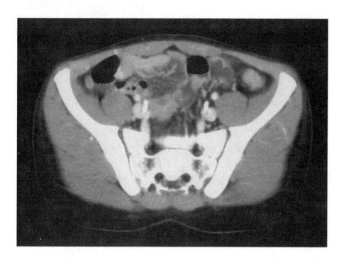
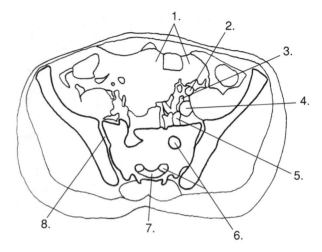

Figure 3-7

1. What number illustrates the left common iliac vein?
   _____ A. 3
   _____ B. 2
   _____ C. 4
   _____ D. 5

2. Which of the following is illustrated by 7?
   _____ A. Left sacral foramen
   _____ B. Left internal iliac artery
   _____ C. Sacral canal
   _____ D. Intervertebral foramen

3. What number illustrates the left sacral foramina?
   _____ A. 8
   _____ B. 7
   _____ C. 6
   _____ D. 5

4. Which of the following is illustrated by 5?
   _____ A. Left internal iliac artery
   _____ B. Left external iliac artery
   _____ C. Left common iliac vein
   _____ D. Left ureter

5. Which of the following is illustrated by 3?
   _____ A. Left external iliac artery
   _____ B. Left ureter
   _____ C. Left external iliac vein
   _____ D. Left internal iliac artery

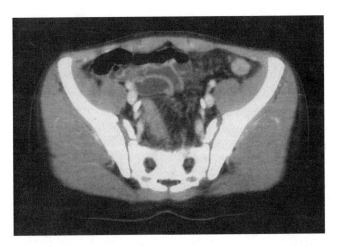 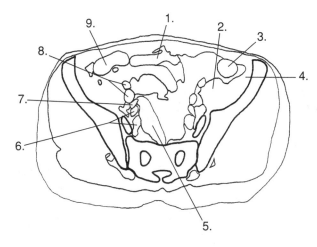

Figure 3-8

1. Which of the following is illustrated by 5?

    _____ A. Small bowel

    _____ B. Descending colon

    _____ C. Top of sigmoid colon

    _____ D. Top of bladder

2. What number illustrates the external iliac artery and vein?

    _____ A. 9

    _____ B. 7

    _____ C. 8

    _____ D. 6

3. Which of the following is illustrated by 6?

    _____ A. Right common iliac artery and vein

    _____ B. Right external iliac artery and vein

    _____ C. Left external iliac artery and vein

    _____ D. Internal iliac artery and vein

4. What number illustrates the iliacus muscle?

    _____ A. 3

    _____ B. 1

    _____ C. 2

    _____ D. 4

5. What number illustrates the ureter?

    _____ A. 9

    _____ B. 8

    _____ C. 7

    _____ D. 6

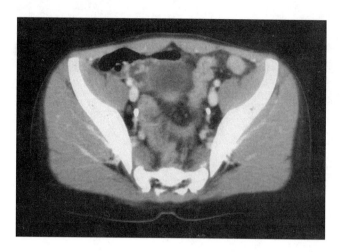
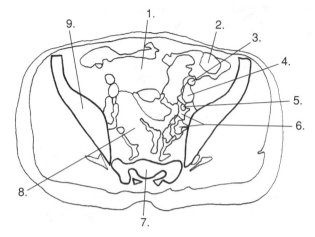

Figure 3-9

1. What number illustrates the descending colon?

   _____ A. 8

   _____ B. 2

   _____ C. 1

   _____ D. 3

2. Which of the following is illustrated by 1?

   _____ A. Top of sigmoid colon

   _____ B. Top of bladder

   _____ C. Sacrum

   _____ D. Descending colon

3. Which of the following is illustrated by 4?

   _____ A. Left external iliac artery

   _____ B. Left ureter

   _____ C. Left common iliac artery

   _____ D. Left external iliac vein

4. What number illustrates the left external iliac artery?

   _____ A. 4

   _____ B. 3

   _____ C. 5

   _____ D. 2

5. Which of the following is illustrated by 5?

   _____ A. Left ureter

   _____ B. Left external iliac artery

   _____ C. Left common iliac artery

   _____ D. Left external iliac vein

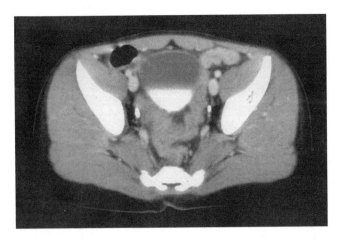 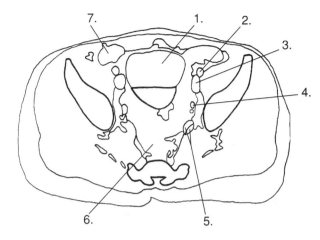

Figure 3-10

1. What number illustrates the left external iliac artery?

_____ A. 2

_____ B. 5

_____ C. 4

_____ D. 3

2. Which of the following is illustrated by 6?

_____ A. Bladder

_____ B. Uterus

_____ C. Sigmoid colon

_____ D. Descending colon

3. What number illustrates the left ureter?

_____ A. 2

_____ B. 5

_____ C. 4

_____ D. 3

4. Which of the following is illustrated by 1?

_____ A. Small bowel

_____ B. Bladder

_____ C. Sigmoid colon

_____ D. Descending colon

5. Which of the following is illustrated by 3?

_____ A. Small bowel

_____ B. Left ureter

_____ C. Left external iliac artery

_____ D. Left external iliac vein

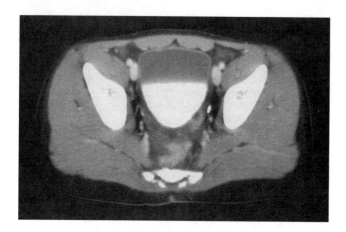
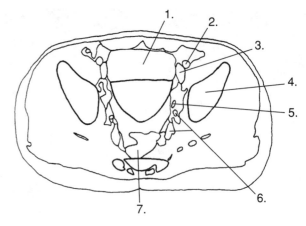

Figure 3-11

1. Which of the following is illustrated by 5?

    _____ A. Internal iliac vessels

    _____ B. External iliac vein

    _____ C. Ureter

    _____ D. External iliac artery

2. What number illustrates the left external iliac artery?

    _____ A. 2

    _____ B. 5

    _____ C. 3

    _____ D. 6

3. Which of the following is illustrated by 7?

    _____ A. Rectum

    _____ B. Sigmoid colon

    _____ C. Descending colon

    _____ D. Seminal vesicles

4. Which of the following is illustrated by 4?

    _____ A. Pubic bone

    _____ B. Iliac bone

    _____ C. Ischial bone

    _____ D. Proximal femur

5. Which of the following is illustrated by 3?

    _____ A. Femoral artery

    _____ B. Ureter

    _____ C. External iliac artery

    _____ D. External iliac vein

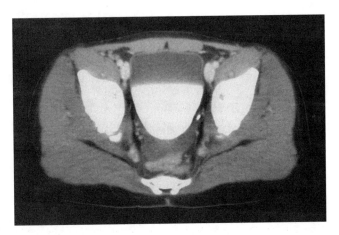 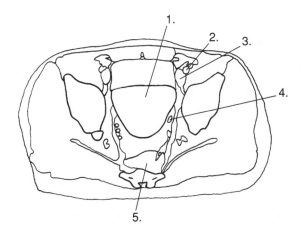

Figure 3-12

1. What number illustrates the left external iliac vein?

     _____ A. 5

     _____ B. 3

     _____ C. 2

     _____ D. 4

2. Which of the following is illustrated by 5?

     _____ A. Rectum

     _____ B. Seminal vesicles

     _____ C. Sigmoid colon

     _____ D. Prostate gland

3. Which of the following is illustrated by 4?

     _____ A. Left seminal vesicle

     _____ B. Left ureter

     _____ C. Left external iliac artery

     _____ D. Left external iliac vein

4. What number illustrates the bladder?

     _____ A. 1

     _____ B. 5

     _____ C. 4

     _____ D. 2

5. What number illustrates the left external iliac artery?

     _____ A. 4

     _____ B. 2

     _____ C. 3

     _____ D. 1

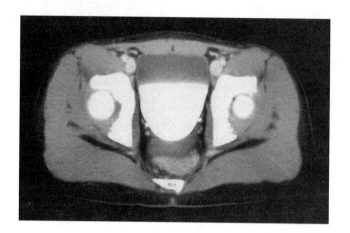 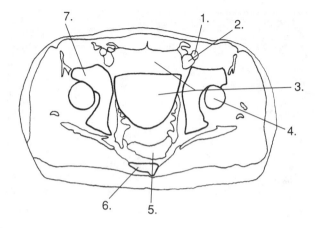

Figure 3-13

1. What number illustrates the tip of the sacrum?

    _____ A. 2

    _____ B. 5

    _____ C. 4

    _____ D. 6

2. Which of the following is illustrated by 1?

    _____ A. External iliac vein

    _____ B. External iliac artery

    _____ C. Spermatic cord

    _____ D. Femoral artery

3. Which of the following is illustrated by 7?

    _____ A. Right ischium

    _____ B. Right ilium

    _____ C. Right pubis

    _____ D. Greater trochanter of femur

4. What number illustrates the head of the femur?

    _____ A. 7

    _____ B. 4

    _____ C. 6

    _____ D. 1

5. What number illustrates the external iliac artery?

    _____ A. 1

    _____ B. 2

    _____ C. 7

    _____ D. 5

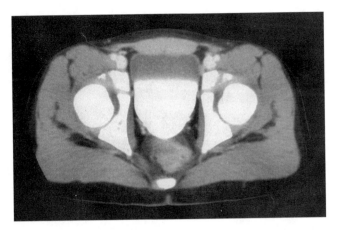 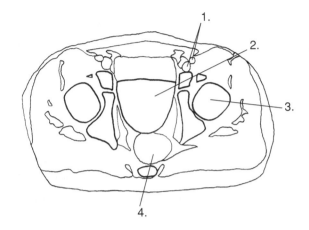

Figure 3-14

1. What number illustrates the rectum?

   _____ A. 3

   _____ B. 2

   _____ C. 1

   _____ D. 4

2. Which of the following is illustrated by 3?

   _____ A. Head of right femur

   _____ B. Head of left femur

   _____ C. Right ilium

   _____ D. Left ilium

3. What number illustrates the bladder?

   _____ A. 3

   _____ B. 2

   _____ C. 4

   _____ D. 1

4. Which of the following is illustrated by 1?

   _____ A. Right exterior iliac artery and vein

   _____ B. Left exterior iliac artery and vein

   _____ C. Right femoral artery and vein

   _____ D. Left femoral artery and vein

5. Which of the following is illustrated by 4?

   _____ A. Seminal vesicles

   _____ B. Sigmoid colon

   _____ C. Rectum

   _____ D. Bladder

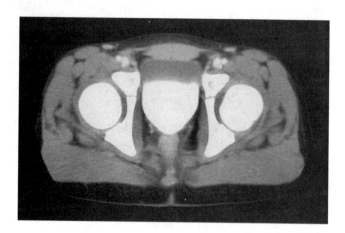 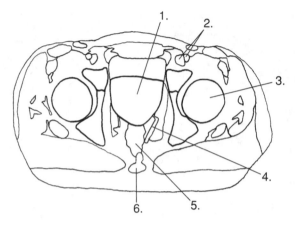

Figure 3-15

1. Which of the following is illustrated by 2?

_____ A. Right external iliac artery and vein

_____ B. Left femoral artery and vein

_____ C. Left external iliac artery and vein

_____ D. Right femoral artery and vein

2. What number illustrates the coccyx?

_____ A. 5

_____ B. 2

_____ C. 1

_____ D. 6

3. What number illustrates the rectum?

_____ A. 1

_____ B. 6

_____ C. 5

_____ D. 3

4. Which of the following is illustrated by 1?

_____ A. Coccyx

_____ B. Head of the left femur

_____ C. Bladder

_____ D. Rectum

5. What number illustrates the pelvic diaphragm?

_____ A. 6

_____ B. 5

_____ C. 4

_____ D. 1

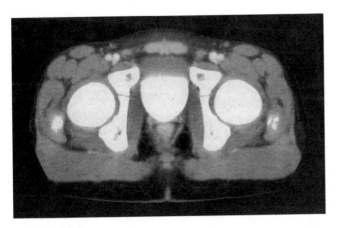
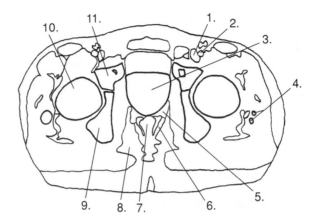

Figure 3-16

1. Which of the following is illustrated by 9?

_____ A. Right pubis

_____ B. Right ileum

_____ C. Right ischium

_____ D. Right ilium

2. What number illustrates the pelvic diaphragm?

_____ A. 5

_____ B. 7

_____ C. 8

_____ D. 6

3. Which of the following is illustrated by 1?

_____ A. Femoral vein

_____ B. Femoral artery

_____ C. External iliac artery

_____ D. External iliac vein

4. What number illustrates the pubis?

_____ A. 10

_____ B. 9

_____ C. 11

_____ D. 3

5. What number illustrates the seminal vesicle?

_____ A. 7

_____ B. 6

_____ C. 5

_____ D. 8

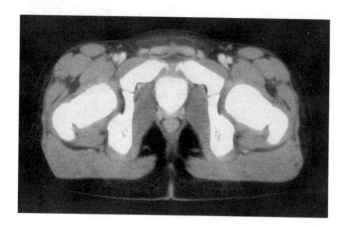 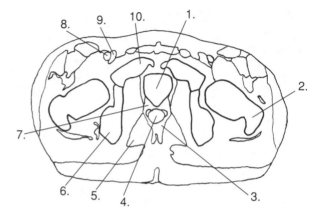

Figure 3-17

1. What number illustrates the right femoral vein?

    _____ A. 8

    _____ B. 6

    _____ C. 4

    _____ D. 9

2. Which of the following is illustrated by 7?

    _____ A. Seminal vesicle

    _____ B. Adnexal area

    _____ C. Prostate

    _____ D. Ischiorectal fossa

3. What number illustrates the bladder?

    _____ A. 1

    _____ B. 7

    _____ C. 3

    _____ D. 4

4. Which of the following is illustrated by 6?

    _____ A. Pubis

    _____ B. Ileum

    _____ C. Ischium

    _____ D. Ilium

5. Which of the following is illustrated by 4?

    _____ A. Pelvic diaphragm

    _____ B. Rectum

    _____ C. Bladder

    _____ D. Sigmoid colon

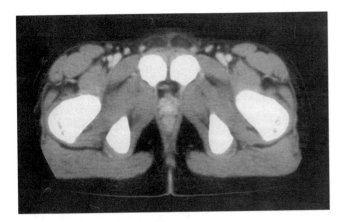
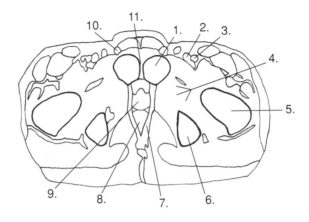

Figure 3-18

1. Which of the following is illustrated by 6?

_____ A. Ischial tuberosity

_____ B. Pubic bone

_____ C. Ischial spine

_____ D. Proximal femur

2. Which of the following is illustrated by 1?

_____ A. Ilium

_____ B. Pubic bone

_____ C. Ischial spine

_____ D. Ischial tuberosity

3. What number illustrates the prostate?

_____ A. 6

_____ B. 8

_____ C. 1

_____ D. 9

4. What number illustrates the femoral artery?

_____ A. 10

_____ B. 2

_____ C. 3

_____ D. 1

5. Which of the following is illustrated by 5?

_____ A. Neck of the left femur

_____ B. Greater trochanter of femur

_____ C. Lesser trochanter of femur

_____ D. Ischial bone

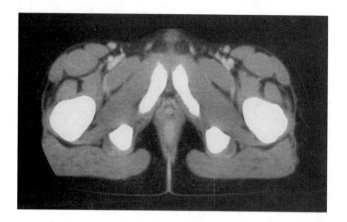 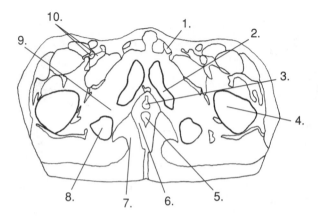

Figure 3-19

1. Which of the following is illustrated by 6?

   _____ A. Pelvic diaphragm

   _____ B. Obturator foramina

   _____ C. Ischiorectal fossa

   _____ D. Prostate

2. What number illustrates the ischial tuberosity?

   _____ A. 9

   _____ B. 2

   _____ C. 8

   _____ D. 4

3. What number illustrates the spermatic cord?

   _____ A. 1

   _____ B. 5

   _____ C. 10

   _____ D. 3

4. Which of the following is illustrated by 2?

   _____ A. Superior pubic ramus

   _____ B. Inferior pubic ramus

   _____ C. Ischial tuberosity

   _____ D. Ischial ramus

5. What number illustrates the prostate?

   _____ A. 6

   _____ B. 3

   _____ C. 5

   _____ D. 2

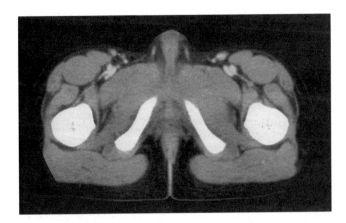 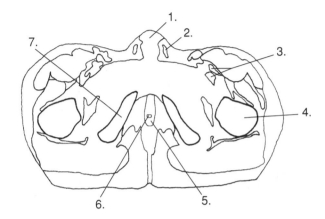

Figure 3-20

1. Which of the following is illustrated by 1?

      _____ A. Corpus spongiosum

      _____ B. Corpus cavernosum

      _____ C. Urethra

      _____ D. Symphysis pubis

2. What number illustrates the urethra?

      _____ A. 5

      _____ B. 2

      _____ C. 6

      _____ D. 1

3. What number illustrates the ischial ramus?

      _____ A. 5

      _____ B. 2

      _____ C. 1

      _____ D. 7

4. Which of the following is illustrated by 2?

      _____ A. Femoral artery

      _____ B. Femoral vein

      _____ C. Spermatic cord

      _____ D. Urethra

5. Which of the following is illustrated by 5?

      _____ A. Corpus cavernosum

      _____ B. Left spermatic cord

      _____ C. Urethra

      _____ D. Corpus spongiosum

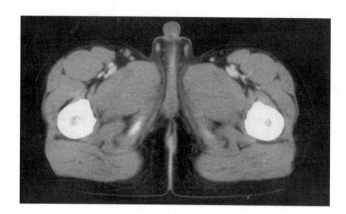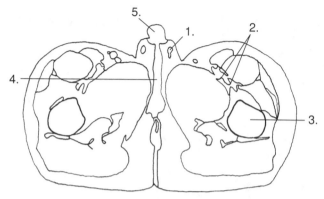

Figure 3-21

1. What number illustrates the corpus spongiosum?

    _____ A. 5

    _____ B. 2

    _____ C. 4

    _____ D. 1

2. Which of the following is illustrated by 1?

    _____ A. Femoral artery

    _____ B. Spermatic cord

    _____ C. Femoral vein

    _____ D. External iliac artery

3. What number illustrates the corpus cavernosum?

    _____ A. 3

    _____ B. 1

    _____ C. 4

    _____ D. 5

4. Which of the following is illustrated by 2?

    _____ A. Femoral vessels

    _____ B. Corpus spongiosum

    _____ C. Spermatic cord

    _____ D. External iliac vessels

5. Which of the following is illustrated by 3?

    _____ A. Ischial tuberosity

    _____ B. Head of femur

    _____ C. Shaft of the left femur

    _____ D. Greater trochanter of femur

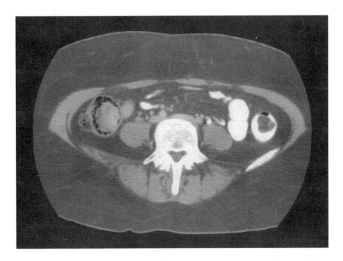 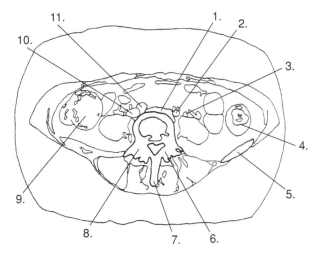

Figure 3-22

1. What number illustrates the pedicle of L5?

_____ A. 8

_____ B. 6

_____ C. 7

_____ D. 9

2. Which of the following is illustrated by 2?

_____ A. Ureter

_____ B. Internal iliac vein

_____ C. Common iliac artery

_____ D. Common iliac vein

3. What number illustrates the left common iliac vein?

_____ A. 11

_____ B. 2

_____ C. 1

_____ D. 3

4. Which of the following is illustrated by 6?

_____ A. Intervertebral foramen

_____ B. Lamina of L5

_____ C. Spinous process of L5

_____ D. Pedicle of L5

5. Which of the following is illustrated by 3?

_____ A. Common iliac vein

_____ B. Common iliac artery

_____ C. External iliac artery

_____ D. Ureter

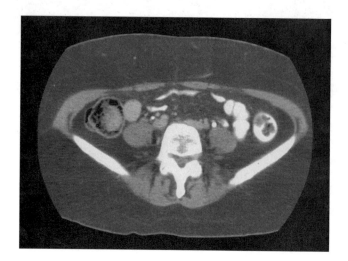
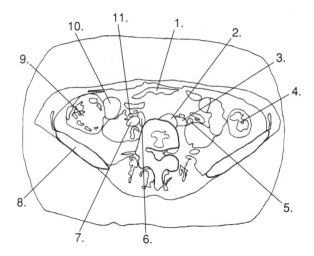

Figure 3-23

1. Which of the following is illustrated by 4?

    _____ A. Descending colon

    _____ B. Ascending colon

    _____ C. Ileum

    _____ D. Cecum

2. What number illustrates the left common iliac artery?

    _____ A. 5

    _____ B. 2

    _____ C. 3

    _____ D. 6

3. Which of the following is illustrated by 1?

    _____ A. Cecum

    _____ B. Transverse colon

    _____ C. Small bowel

    _____ D. Sigmoid colon

4. Which number illustrates the right common iliac vein?

    _____ A. 10

    _____ B. 7

    _____ C. 11

    _____ D. 6

5. Which of the following is illustrated by 11?

    _____ A. Ureter

    _____ B. Common iliac vein

    _____ C. External iliac artery

    _____ D. Common iliac artery

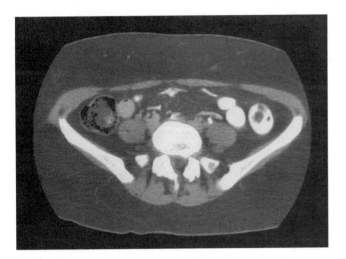 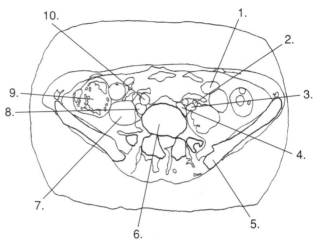

Figure 3-24

1. Which of the following is illustrated by 2?

    _____ A. Common iliac vein

    _____ B. Ureter

    _____ C. Common iliac artery

    _____ D. External iliac artery

2. What number illustrates the right common iliac artery?

    _____ A. 3

    _____ B. 8

    _____ C. 2

    _____ D. 10

3. What number illustrates the left common iliac artery?

    _____ A. 2

    _____ B. 3

    _____ C. 4

    _____ D. 10

4. What number illustrates the left common iliac vein?

    _____ A. 3

    _____ B. 2

    _____ C. 1

    _____ D. 4

5. Which of the following is illustrated by 10?

    _____ A. Common iliac vein

    _____ B. Psoas muscle

    _____ C. Common iliac artery

    _____ D. Inferior mesenteric artery

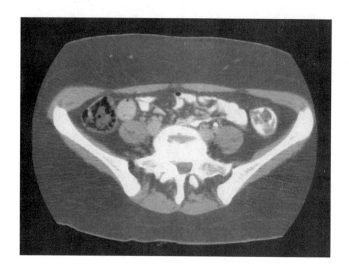 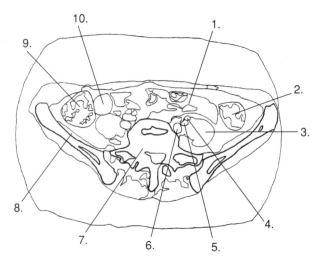

Figure 3-25

1. What number illustrates the small bowel?

    _____ A. 1

    _____ B. 3

    _____ C. 2

    _____ D. 9

2. Which of the following is illustrated by 4?

    _____ A. Common iliac artery

    _____ B. Ureter

    _____ C. Common iliac vein

    _____ D. External iliac artery

3. Which of the following is illustrated by 8?

    _____ A. Psoas muscle

    _____ B. Iliacus muscle

    _____ C. Ileum

    _____ D. Ilium

4. What number illustrates the common iliac artery?

    _____ A. 6

    _____ B. 3

    _____ C. 5

    _____ D. 4

5. What number illustrates the descending colon?

    _____ A. 2

    _____ B. 9

    _____ C. 10

    _____ D. 3

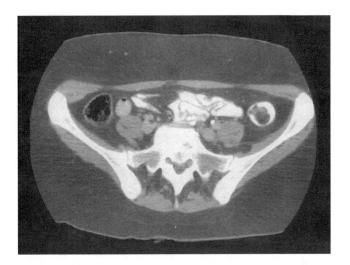 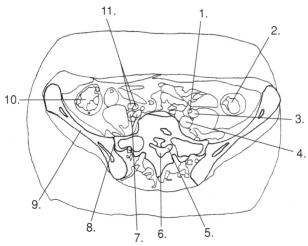

Figure 3-26

1. What number illustrates the right sacroiliac joint?

    _____ A. 9

    _____ B. 5

    _____ C. 8

    _____ D. 6

2. Which of the following is illustrated by 4?

    _____ A. Common iliac vein

    _____ B. Ureter

    _____ C. Internal iliac artery

    _____ D. Common iliac artery

3. Which of the following is illustrated by 3?

    _____ A. Internal iliac artery

    _____ B. Common iliac artery

    _____ C. Ureter

    _____ D. External iliac vein

4. What number illustrates the cecum?

    _____ A. 2

    _____ B. 9

    _____ C. 4

    _____ D. 10

5. What number illustrates the descending colon?

    _____ A. 10

    _____ B. 6

    _____ C. 3

    _____ D. 2

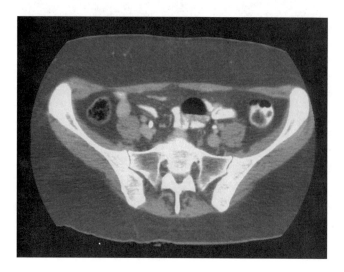
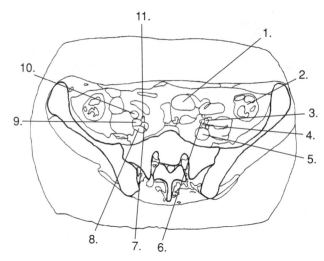

Figure 3-27

1. Which of the following is illustrated by 6?

    _____ A. External iliac artery

    _____ B. Internal iliac artery

    _____ C. Internal iliac vein

    _____ D. Ureter

2. What number illustrates the right external iliac artery?

    _____ A. 11

    _____ B. 8

    _____ C. 10

    _____ D. 7

3. Which of the following illustrates the right internal iliac vein?

    _____ A. 9

    _____ B. 8

    _____ C. 7

    _____ D. 10

4. What number illustrates the left common iliac vein?

    _____ A. 6

    _____ B. 4

    _____ C. 5

    _____ D. 3

5. Which of the following illustrates the left external iliac artery?

    _____ A. 5

    _____ B. 4

    _____ C. 3

    _____ D. 6

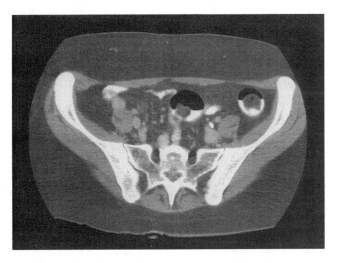
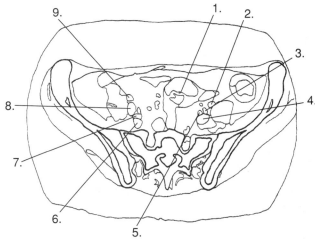

Figure 3-28

1. Which of the following is illustrated by 8?

  _____ A. External iliac vein

  _____ B. Internal iliac vein

  _____ C. External iliac artery

  _____ D. Internal iliac artery

2. What number illustrates the external iliac artery?

  _____ A. 2

  _____ B. 4

  _____ C. 3

  _____ D. 5

3. Which of the following is illustrated by 5?

  _____ A. Common iliac vein

  _____ B. Ureter

  _____ C. Internal iliac artery

  _____ D. External iliac artery

4. What number illustrates the internal iliac artery?

  _____ A. 8

  _____ B. 7

  _____ C. 6

  _____ D. 9

5. Which of the following is illustrated by 3?

  _____ A. Ureter

  _____ B. External iliac artery

  _____ C. Common iliac vein

  _____ D. Internal iliac artery

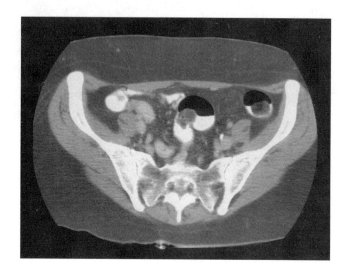 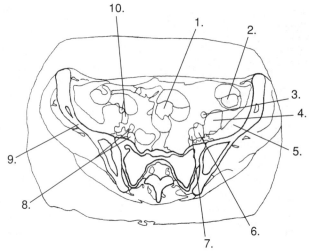

Figure 3-29

1. Which of the following is illustrated by 2?

   _____ A. Descending colon

   _____ B. Sigmoid colon

   _____ C. Left psoas muscle

   _____ D. Ascending colon

2. What number illustrates the left internal iliac artery?

   _____ A. 6

   _____ B. 3

   _____ C. 4

   _____ D. 7

3. What number illustrates the iliacus muscle?

   _____ A. 4

   _____ B. 2

   _____ C. 5

   _____ D. 6

4. Which of the following is illustrated by 4?

   _____ A. Small bowel

   _____ B. Psoas muscle

   _____ C. Internal iliac vein

   _____ D. Sigmoid colon

5. Which of the following is illustrated by 6?

   _____ A. External iliac artery

   _____ B. Internal iliac artery

   _____ C. Internal iliac vein

   _____ D. Common iliac vein

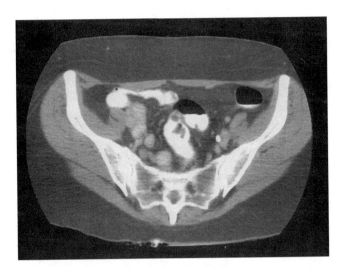 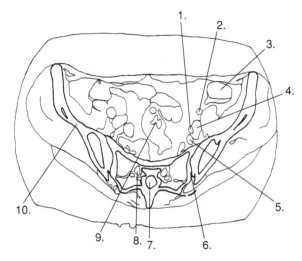

Figure 3-30

1. Which of the following is illustrated by 4?

    _____ A. External iliac artery

    _____ B. Common iliac vein

    _____ C. Ureter

    _____ D. Internal iliac artery

2. What number illustrates the sigmoid colon?

    _____ A. 6

    _____ B. 5

    _____ C. 8

    _____ D. 9

3. Which of the following is illustrated by 5?

    _____ A. Internal iliac artery

    _____ B. Ureter

    _____ C. Common iliac vein

    _____ D. External iliac artery

4. What number illustrates the sacral canal?

    _____ A. 7

    _____ B. 9

    _____ C. 8

    _____ D. 10

5. What number illustrates the sacral foramen?

    _____ A. 10

    _____ B. 7

    _____ C. 8

    _____ D. 6

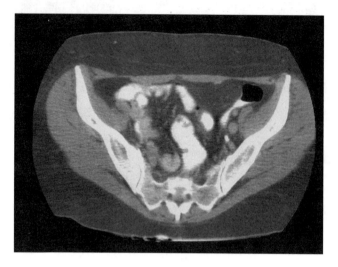 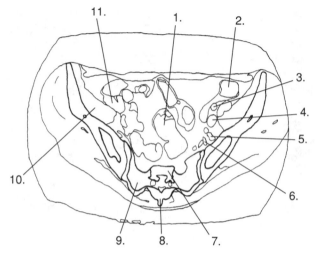

Figure 3-31

1. Which of the following is illustrated by 9?

    _____ A. Lateral part of the sacrum

    _____ B. Spinous process

    _____ C. Vertebral body

    _____ D. Sigmoid colon

2. What number illustrates the psoas muscle?

    _____ A. 10

    _____ B. 9

    _____ C. 1

    _____ D. 11

3. Which of the following is illustrated by 5?

    _____ A. External iliac artery

    _____ B. Internal iliac vein

    _____ C. External iliac vein

    _____ D. Internal iliac artery

4. What number illustrates the spinous process?

    _____ A. 9

    _____ B. 8

    _____ C. 7

    _____ D. 11

5. Which of the following illustrates the iliacus muscle?

    _____ A. 11

    _____ B. 9

    _____ C. 10

    _____ D. 8

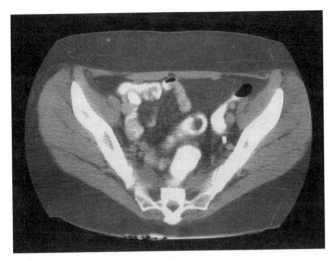 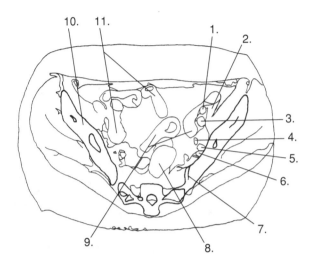

Figure 3-32

1. Which of the following is illustrated by 8?

    _____ A. Ileum

    _____ B. Rectum

    _____ C. Sigmoid colon

    _____ D. Uterus

2. What number illustrates the left internal iliac artery?

    _____ A. 5

    _____ B. 6

    _____ C. 3

    _____ D. 1

3. Which of the following is illustrated by 4?

    _____ A. Internal iliac vein

    _____ B. Internal iliac artery

    _____ C. External iliac vein

    _____ D. Ureter

4. What number illustrates the sigmoid colon?

    _____ A. 8

    _____ B. 7

    _____ C. 9

    _____ D. 11

5. Which of the following is illustrated by 1?

    _____ A. External iliac artery

    _____ B. Internal iliac vein

    _____ C. Internal iliac artery

    _____ D. External iliac vein

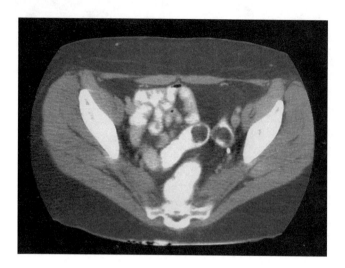
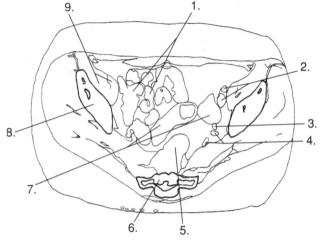

Figure 3-33

1. Which of the following is illustrated by 9?
   _____ A. Loops of the ileum
   _____ B. Iliac muscle
   _____ C. Iliopsoas muscle
   _____ D. Psoas muscle

2. What number illustrates the rectum?
   _____ A. 5
   _____ B. 9
   _____ C. 1
   _____ D. 7

3. Which of the following is illustrated by 6?
   _____ A. Ilium
   _____ B. Ischium
   _____ C. Coccyx
   _____ D. Sacrum

4. Which of the following is illustrated by 7?
   _____ A. Descending colon
   _____ B. Rectum
   _____ C. Loops of the ileum
   _____ D. Sigmoid colon

5. What number illustrates the ilium?
   _____ A. 5
   _____ B. 1
   _____ C. 8
   _____ D. 9

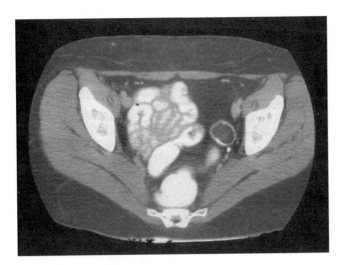 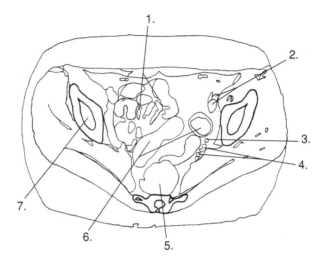

Figure 3-34

1. Which of the following is illustrated by 2?

    _____ A. Femoral artery and vein

    _____ B. External iliac artery and vein

    _____ C. Gluteal artery and vein

    _____ D. Internal iliac artery and vein

2. What number illustrates the sigmoid colon?

    _____ A. 4

    _____ B. 1

    _____ C. 5

    _____ D. 6

3. What number illustrates the left internal iliac artery and vein?

    _____ A. 2

    _____ B. 4

    _____ C. 3

    _____ D. 6

4. Which of the following is illustrated by 1?

    _____ A. Sigmoid colon

    _____ B. Top of bladder

    _____ C. Loops of ileum

    _____ D. Loops of ilium

5. What number illustrates the rectum?

    _____ A. 7

    _____ B. 5

    _____ C. 6

    _____ D. 3

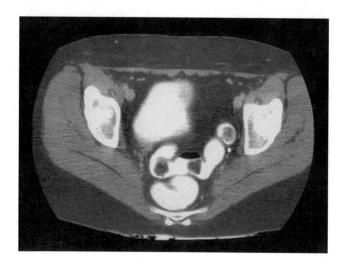 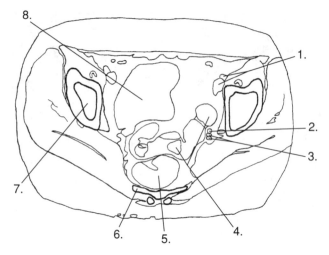

Figure 3-35

1. Which of the following is illustrated by 7?

_____ A. Ischium

_____ B. Pubis

_____ C. Ilium

_____ D. Ileum

2. Which of the following is illustrated by 4?

_____ A. Small bowel

_____ B. Sigmoid colon

_____ C. Descending colon

_____ D. Rectum

3. What number illustrates the rectum?

_____ A. 6

_____ B. 5

_____ C. 4

_____ D. 8

4. Which of the following illustrates the bladder?

_____ A. 5

_____ B. 4

_____ C. 7

_____ D. 8

5. Which of the following is illustrated by 3?

_____ A. Gluteal artery and vein

_____ B. Common iliac artery and vein

_____ C. External iliac artery and vein

_____ D. Internal iliac artery and vein

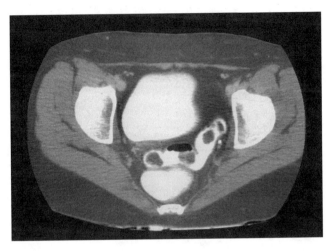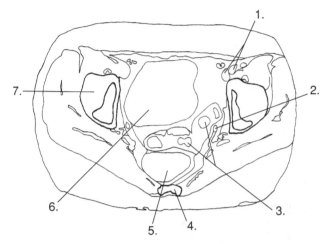

Figure 3-36

1. Which of the following is illustrated by 7?

     _____ A. Ileum

     _____ B. Ilium

     _____ C. Ischium

     _____ D. Pubis

2. Which of the following is illustrated by 2?

     _____ A. External iliac artery

     _____ B. Internal iliac artery

     _____ C. Ureter

     _____ D. Internal iliac vein

3. Which of the following is illustrated by 3?

     _____ A. Sigmoid colon

     _____ B. Rectum

     _____ C. Small bowel

     _____ D. Descending colon

4. What number illustrates the rectum?

     _____ A. 2

     _____ B. 6

     _____ C. 5

     _____ D. 3

5. What number illustrates the bladder?

     _____ A. 5

     _____ B. 6

     _____ C. 3

     _____ D. 2

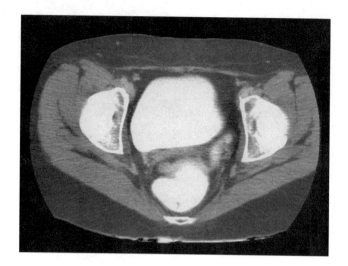
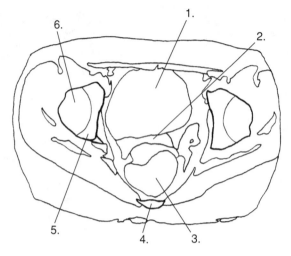

Figure 3-37

1. Which of the following is illustrated by 2?
   _____ A. Cervix of uterus
   _____ B. Vagina
   _____ C. Adnexal area
   _____ D. Fundus of uterus

2. Which of the following is illustrated by 4?
   _____ A. Sacrum
   _____ B. Coccyx
   _____ C. Ischium
   _____ D. Pubis

3. Which of the following is illustrated by 5?
   _____ A. Ilium
   _____ B. Ischium
   _____ C. Pubis
   _____ D. Femur

4. Which of the following is illustrated by 3?
   _____ A. Sigmoid colon
   _____ B. Rectum
   _____ C. Vagina
   _____ D. Cervix of uterus

5. Which of the following is illustrated by 6?
   _____ A. Shaft of femur
   _____ B. Ilium
   _____ C. Neck of femur
   _____ D. Head of femur

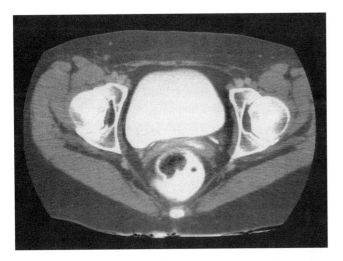

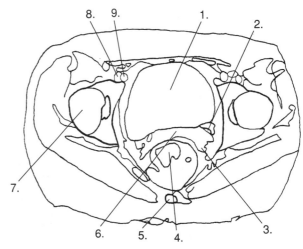

Figure 3-38

1. Which of the following is illustrated by 3?

    _____ A. Adnexal area

    _____ B. Pelvic diaphragm

    _____ C. Body of uterus

    _____ D. Seminal vesicle

2. What number illustrates the femoral artery?

    _____ A. 2

    _____ B. 9

    _____ C. 8

    _____ D. 5

3. What number illustrates the femoral vein?

    _____ A. 3

    _____ B. 8

    _____ C. 9

    _____ D. 7

4. Which of the following is illustrated by 2?

    _____ A. Internal iliac artery

    _____ B. Ureter

    _____ C. Femoral nerve

    _____ D. Internal iliac vein

5. Which of the following is illustrated by 5?

    _____ A. Ilium

    _____ B. Ischium

    _____ C. Sacrum

    _____ D. Coccyx

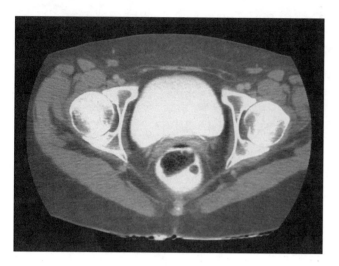 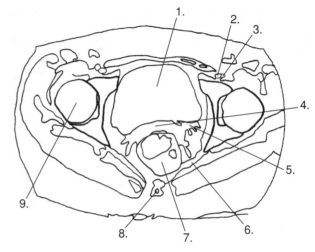

Figure 3-39

1. What number illustrates the femoral vein?

    _____ A. 8

    _____ B. 2

    _____ C. 3

    _____ D. 9

2. Which of the following is illustrated by 7?

    _____ A. Left adnexal area

    _____ B. Body of the uterus

    _____ C. Bladder

    _____ D. Rectum

3. Which of the following is illustrated by 3?

    _____ A. Femoral vein

    _____ B. Femoral artery

    _____ C. External iliac artery

    _____ D. External iliac vein

4. What number illustrates the pelvic diaphragm?

    _____ A. 4

    _____ B. 7

    _____ C. 6

    _____ D. 5

5. Which of the following is illustrated by 5?

    _____ A. Adnexal area

    _____ B. Body of uterus

    _____ C. Pelvic diaphragm

    _____ D. Ureter

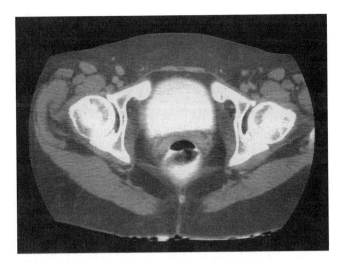 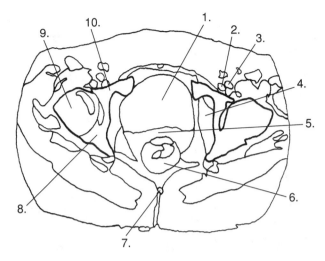

Figure 3-40

1. Which of the following is illustrated by 4?

_____ A. Left adnexal area

_____ B. Body of uterus

_____ C. Pelvic diaphragm

_____ D. Obturator foramen

2. What number illustrates the pubis?

_____ A. 5

_____ B. 9

_____ C. 8

_____ D. 10

3. Which of the following is illustrated by 9?

_____ A. Right pubis

_____ B. Head of right femur

_____ C. Right ischium

_____ D. Greater trochanter of femur

4. Which of the following is illustrated by 7?

_____ A. Coccyx

_____ B. Sacrum

_____ C. Ischium

_____ D. Pubis

5. What number illustrates the ischium?

_____ A. 4

_____ B. 9

_____ C. 8

_____ D. 5

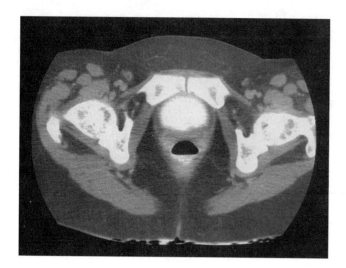 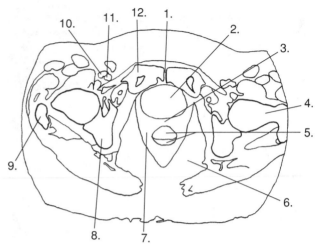

Figure 3-41

1. What number illustrates the pubis?

    _____ A. 9

    _____ B. 8

    _____ C. 1

    _____ D. 12

2. Which of the following is illustrated by 1?

    _____ A. Right pubis

    _____ B. Left ischiorectal fossa

    _____ C. Symphysis pubis

    _____ D. Obturator foramen

3. What number illustrates the pelvic diaphragm?

    _____ A. 1

    _____ B. 6

    _____ C. 3

    _____ D. 7

4. Which of the following is illustrated by 3?

    _____ A. Cervix of uterus

    _____ B. Body of uterus

    _____ C. Pelvic diaphragm

    _____ D. Left ischiorectal fossa

5. What number illustrates the ischiorectal fossa?

    _____ A. 4

    _____ B. 8

    _____ C. 7

    _____ D. 6

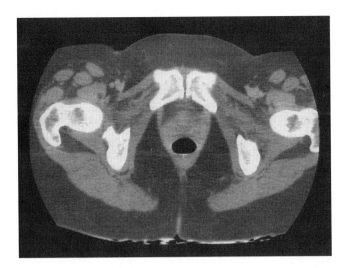
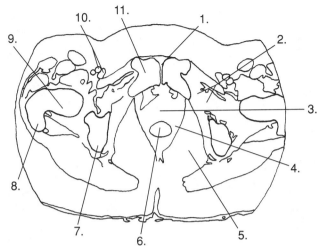

Figure 3-42

1. Which of the following is illustrated by 2?

    _____ A. Symphysis pubis

    _____ B. Pelvic diaphragm

    _____ C. Obturator foramen

    _____ D. Ischiorectal fossa

2. What number illustrates the ischial tuberosity?

    _____ A. 9

    _____ B. 7

    _____ C. 8

    _____ D. 6

3. What number illustrates the ischiorectal fossa?

    _____ A. 2

    _____ B. 4

    _____ C. 3

    _____ D. 5

4. Which of the following is illustrated by 8?

    _____ A. Greater trochanter of right femur

    _____ B. Neck of right femur

    _____ C. Lesser trochanter of right femur

    _____ D. Right ischial tuberosity

5. Which of the following is illustrated by 3?

    _____ A. Body of uterus

    _____ B. Adnexal area

    _____ C. Cervix of uterus

    _____ D. Pelvic diaphragm

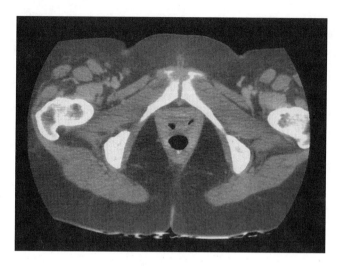 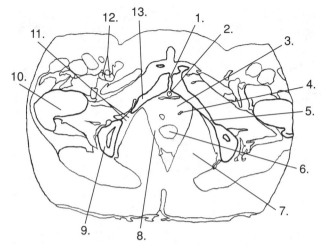

Figure 3-43

1. What number illustrates the cervix of the uterus?

    _____ A. 4

    _____ B. 3

    _____ C. 2

    _____ D. 6

2. Which of the following is illustrated by 2?

    _____ A. Urethra

    _____ B. Post vaginal fornix

    _____ C. Ureter

    _____ D. Cervix of uterus

3. What number illustrates the obturator foramen?

    _____ A. 1

    _____ B. 2

    _____ C. 11

    _____ D. 7

4. Which of the following is illustrated by 5?

    _____ A. Ischial ramus

    _____ B. Ischial tuberosity

    _____ C. Superior pubic ramus

    _____ D. Inferior pubic ramus

5. Which of the following is illustrated by 4?

    _____ A. Symphysis pubis

    _____ B. Post vaginal fornix

    _____ C. Pelvic diaphragm

    _____ D. Urethra

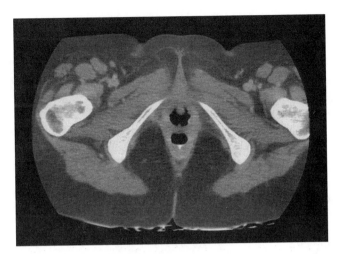 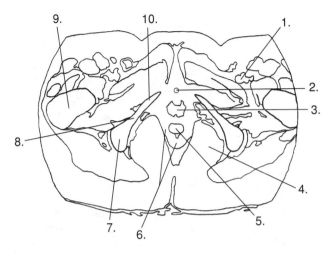

Figure 3-44

1. Which of the following is illustrated by 1?

_____ A. Left ischiorectal fossa

_____ B. Left adnexal area

_____ C. Contents of femoral sheath

_____ D. External iliac vessels

2. Which of the following is illustrated by 9?

_____ A. Ischial ramus

_____ B. Pubic ramus

_____ C. Shaft of femur

_____ D. Ischial tuberosity

3. Which of the following is illustrated by 3?

_____ A. Vagina

_____ B. Rectum

_____ C. Urethra

_____ D. Opening within uterus

4. Which of the following is illustrated by 8?

_____ A. Ilium

_____ B. Ischial tuberosity

_____ C. Pubic ramus

_____ D. Ischial ramus

5. What number illustrates the ischial tuberosity?

_____ A. 7

_____ B. 8

_____ C. 10

_____ D. 9

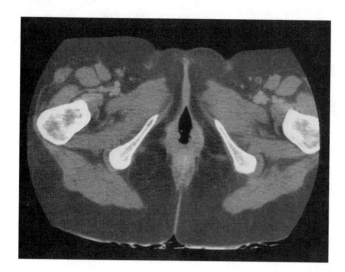 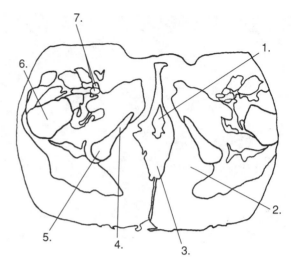

Figure 3-45

1. Which of the following is illustrated by 2?

    _____ A. Pelvic diaphragm

    _____ B. Ischiorectal fossa

    _____ C. Ischial tuberosity

    _____ D. Femoral sheath

2. What number illustrates the ischial tuberosity?

    _____ A. 4

    _____ B. 1

    _____ C. 5

    _____ D. 6

3. Which of the following is illustrated by 1?

    _____ A. Vagina

    _____ B. Rectum

    _____ C. Opening within uterus

    _____ D. Ischiorectal fossa

4. What number illustrates the pelvic diaphragm?

    _____ A. 2

    _____ B. 3

    _____ C. 6

    _____ D. 5

5. What number illustrates the ischial ramus?

    _____ A. 4

    _____ B. 5

    _____ C. 7

    _____ D. 6

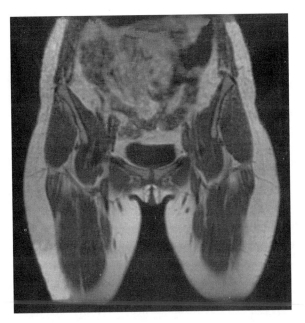
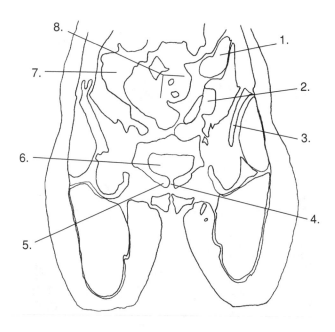

Figure 3-46

1. Which of the following is illustrated by 5?

_____ A. Ilium

_____ B. Ischium

_____ C. Pubis

_____ D. Ileum

2. Which of the following is illustrated by 7?

_____ A. Cecum

_____ B. Ascending colon

_____ C. Descending colon

_____ D. Sigmoid colon

3. Which of the following is illustrated by 2?

_____ A. Cecum

_____ B. Ascending colon

_____ C. Descending colon

_____ D. Sigmoid colon

4. What number illustrates the descending colon?

_____ A. 7

_____ B. 2

_____ C. 1

_____ D. 8

5. What number illustrates the ilium?

_____ A. 5

_____ B. 4

_____ C. 3

_____ D. 8

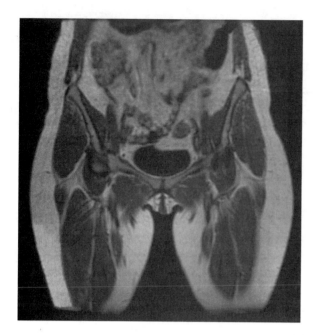 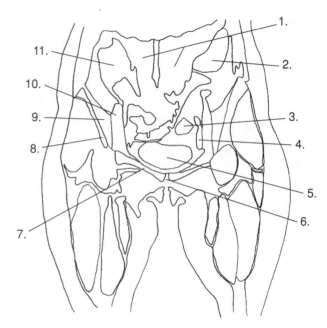

Figure 3-47

1. Which of the following is illustrated by 2?

    _____ A. Cecum

    _____ B. Descending colon

    _____ C. Sigmoid colon

    _____ D. Small bowel

2. Which of the following is illustrated by 3?

    _____ A. Sigmoid colon

    _____ B. Small bowel

    _____ C. Descending colon

    _____ D. Cecum

3. What number illustrates the psoas muscle?

    _____ A. 4

    _____ B. 8

    _____ C. 9

    _____ D. 10

4. Which of the following is illustrated by 4?

    _____ A. Internal iliac artery and vein

    _____ B. External iliac artery and vein

    _____ C. Femoral artery and vein

    _____ D. Common iliac artery and vein

5. Which of the following is illustrated by 1?

    _____ A. Cecum

    _____ B. Descending colon

    _____ C. Sigmoid colon

    _____ D. Small bowel

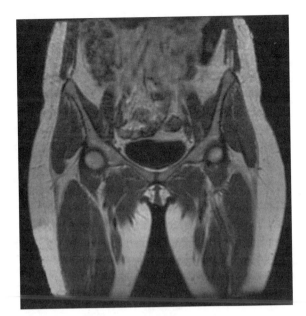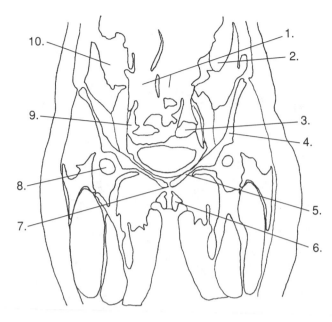

Figure 3-48

1. Which of the following is illustrated by 6?

    _____ A. Labium minora

    _____ B. Pelvic diaphragm

    _____ C. Labium majora

    _____ D. Ischiorectal fossa

2. Which of the following is illustrated by 9?

    _____ A. Internal iliac artery and vein

    _____ B. External iliac artery and vein

    _____ C. Femoral artery and vein

    _____ D. Common iliac artery and vein

3. What number illustrates the descending colon?

    _____ A. 10

    _____ B. 1

    _____ C. 2

    _____ D. 3

4. Which of the following is illustrated by 5?

    _____ A. Ilium

    _____ B. Ischium

    _____ C. Pubis

    _____ D. Pelvic diaphragm

5. What number illustrates the sigmoid colon?

    _____ A. 10

    _____ B. 1

    _____ C. 2

    _____ D. 3

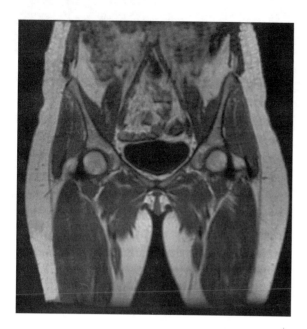 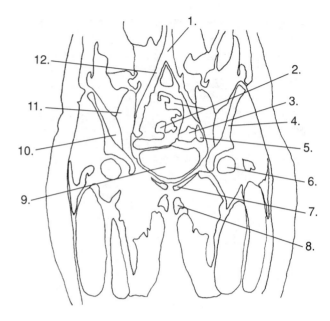

Figure 3-49

1. Which of the following is illustrated by 12?

   _____ A. Internal iliac artery and vein

   _____ B. External iliac artery and vein

   _____ C. Femoral artery and vein

   _____ D. Common iliac artery

2. What number illustrates the small bowel?

   _____ A. 3

   _____ B. 2

   _____ C. 9

   _____ D. 11

3. What number illustrates the iliacus muscle?

   _____ A. 5

   _____ B. 4

   _____ C. 11

   _____ D. 10

4. Which of the following is illustrated by 7?

   _____ A. Ilium

   _____ B. Ischium

   _____ C. Pubis

   _____ D. Pelvic diaphragm

5. Which of the following is illustrated by 5?

   _____ A. Internal iliac artery and vein

   _____ B. External iliac artery and vein

   _____ C. Femoral artery and vein

   _____ D. Common iliac artery and vein

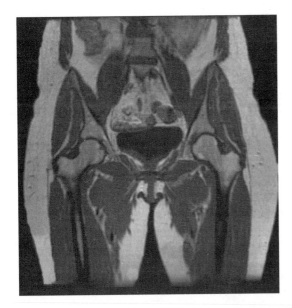 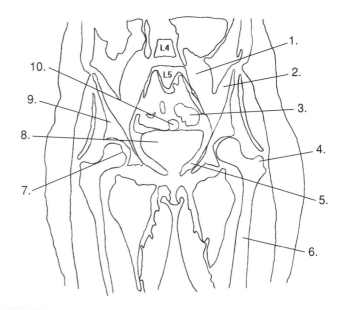

Figure 3-50

1. Which of the following is illustrated by 1?
   _____ A. Descending colon
   _____ B. Sigmoid colon
   _____ C. Psoas muscle
   _____ D. Iliacus muscle

2. Which of the following is illustrated by 7?
   _____ A. Greater trochanter of femur
   _____ B. Shaft of femur
   _____ C. Fovea capitis femoris
   _____ D. Lesser trochanter of femur

3. Which of the following is illustrated by 10?
   _____ A. Small bowel
   _____ B. Sigmoid colon
   _____ C. Cecum
   _____ D. Descending colon

4. What number illustrates the bladder?
   _____ A. 3
   _____ B. 10
   _____ C. 8
   _____ D. 9

5. What number illustrates the ilium?
   _____ A. 3
   _____ B. 10
   _____ C. 9
   _____ D. 8

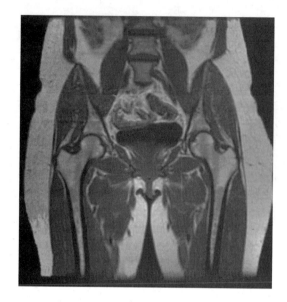 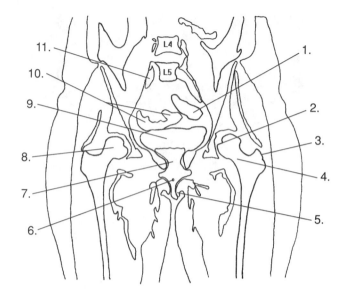

Figure 3-51

1. Which of the following is illustrated by 1?
    _____ A. Small bowel
    _____ B. Sigmoid colon
    _____ C. Psoas muscle
    _____ D. Iliacus muscle

2. Which of the following is illustrated by 11?
    _____ A. Internal iliac artery and vein
    _____ B. External iliac artery and vein
    _____ C. Femoral artery and vein
    _____ D. Common iliac artery and vein

3. Which of the following is illustrated by 7?
    _____ A. Uterus
    _____ B. Wall of bladder
    _____ C. Pelvic diaphragm
    _____ D. Cervix

4. Which of the following is illustrated by 6?
    _____ A. Ureter
    _____ B. Urethra
    _____ C. Vagina
    _____ D. Vaginal fornix

5. What number illustrates the urethra?
    _____ A. 5
    _____ B. 6
    _____ C. 7
    _____ D. 9

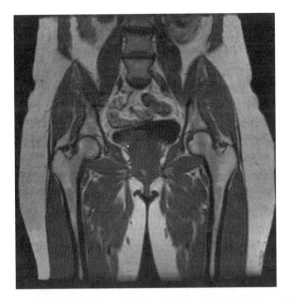 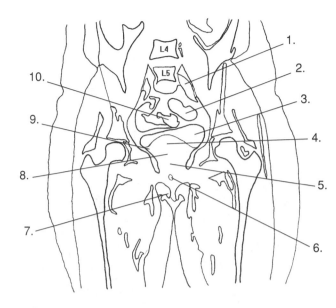

Figure 3-52

1. What number illustrates the adnexal area?

    _____ A. 4

    _____ B. 5

    _____ C. 8

    _____ D. 9

2. What number illustrates the cervix?

    _____ A. 4

    _____ B. 8

    _____ C. 5

    _____ D. 6

3. Which of the following is illustrated by 6?

    _____ A. Urethra

    _____ B. Cervix of uterus

    _____ C. Vagina

    _____ D. Rectum

4. Which of the following is illustrated by 3?

    _____ A. Posterior bladder

    _____ B. Sigmoid colon

    _____ C. Small bowel

    _____ D. Body of uterus

5. Which of the following is illustrated by 4?

    _____ A. Adnexal area

    _____ B. Cervix of uterus

    _____ C. Posterior bladder

    _____ D. Fundus of uterus

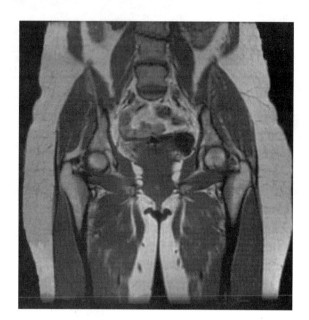 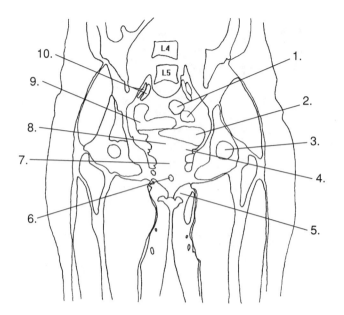

Figure 3-53

1. What number illustrates the sigmoid colon?

    _____ A. 9

    _____ B. 1

    _____ C. 2

    _____ D. 8

2. Which of the following is illustrated by 7?

    _____ A. Urethra

    _____ B. Cervix of uterus

    _____ C. Vagina

    _____ D. Rectum

3. Which of the following is illustrated by 4?

    _____ A. Adnexal area

    _____ B. Cervix of uterus

    _____ C. Posterior bladder

    _____ D. Fundus of uterus

4. Which of the following is illustrated by 3?

    _____ A. Ischium

    _____ B. Femur

    _____ C. Ilium

    _____ D. Pubis

5. Which of the following is illustrated by 10?

    _____ A. Internal iliac artery and vein

    _____ B. External iliac artery and vein

    _____ C. Common iliac artery and vein

    _____ D. Femoral artery and vein

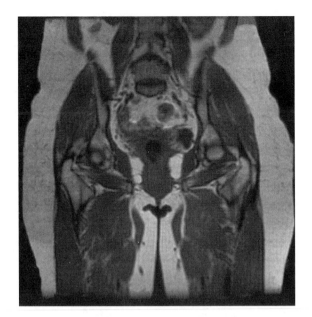
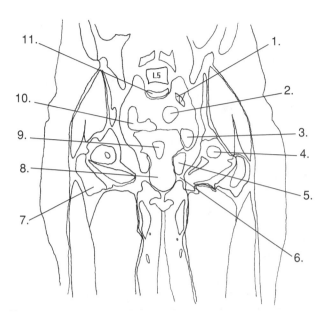

Figure 3-54

1. Which of the following is illustrated by 8?

        _____ A. Vagina

        _____ B. Urethra

        _____ C. Anal sphincter

        _____ D. Ischiorectal fossa

2. Which of the following is illustrated by 9?

        _____ A. Urethra

        _____ B. Vagina

        _____ C. Rectum

        _____ D. Sigmoid colon

3. Which of the following is illustrated by 3?

        _____ A. Small bowel

        _____ B. Sigmoid colon

        _____ C. Adnexal area

        _____ D. Posterior bladder

4. What number illustrates the pelvic diaphragm?

        _____ A. 6

        _____ B. 8

        _____ C. 5

        _____ D. 7

5. Which of the following is illustrated by 1?

        _____ A. Common iliac artery and vein

        _____ B. Internal iliac artery and vein

        _____ C. External iliac artery and vein

        _____ D. Gluteal artery and vein

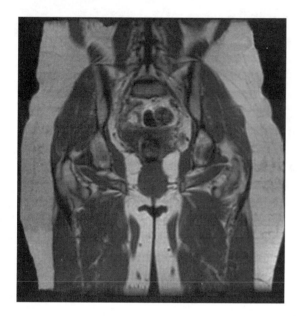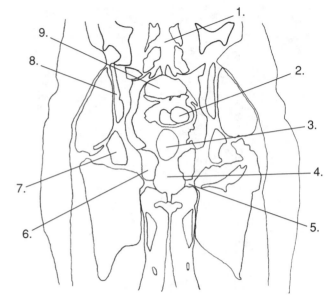

Figure 3-55

1. Which of the following is illustrated by 7?
    _____ A. Ilium
    _____ B. Ischial tuberosity
    _____ C. Pubis
    _____ D. Ischial spine

2. Which of the following is illustrated by 3?
    _____ A. Rectum
    _____ B. Vagina
    _____ C. Sigmoid colon
    _____ D. Small bowel

3. What number illustrates the vertebral foramen?
    _____ A. 1
    _____ B. 9
    _____ C. 2
    _____ D. 3

4. What number illustrates the rectum?
    _____ A. 2
    _____ B. 3
    _____ C. 4
    _____ D. 9

5. What number illustrates the ilium?
    _____ A. 6
    _____ B. 7
    _____ C. 8
    _____ D. 9

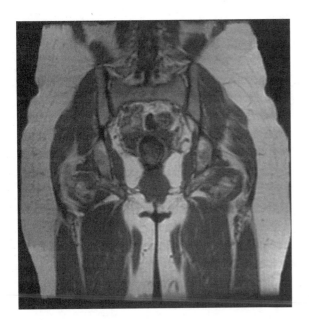 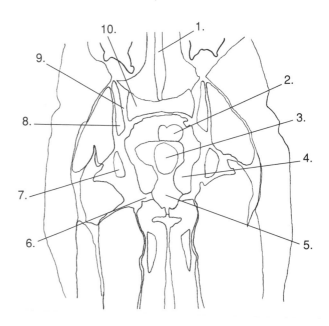

Figure 3-56

1. Which of the following is illustrated by 9?

    _____ A. Ilium

    _____ B. Sacroiliac joint

    _____ C. Sacrum

    _____ D. Ischial tuberosity

2. Which of the following is illustrated by 5?

    _____ A. Rectum

    _____ B. Vagina

    _____ C. Anal sphincter

    _____ D. Pelvic diaphragm

3. What number illustrates the ischial tuberosity?

    _____ A. 9

    _____ B. 8

    _____ C. 7

    _____ D. 6

4. Which of the following is illustrated by 3?

    _____ A. Rectum

    _____ B. Vagina

    _____ C. Sigmoid colon

    _____ D. Anal sphincter

5. What number illustrates the sigmoid colon?

    _____ A. 1

    _____ B. 2

    _____ C. 3

    _____ D. 5

ANSWERS

Fig. 3-1

1. C
2. B
3. C
4. D
5. B

Fig. 3-2

1. C
2. B
3. A
4. D
5. B

Fig. 3-3

1. D
2. B
3. D
4. D
5. D

Fig. 3-4

1. D
2. B
3. C
4. A
5. B

Fig. 3-5

1. C
2. C
3. B
4. B
5. C

Fig. 3-6

1. B
2. C
3. A
4. D
5. B

Fig. 3-7

1. C
2. C
3. C
4. A
5. B

Fig. 3-8

1. C
2. C
3. D
4. D
5. C

Fig. 3-9

1. B
2. B
3. D
4. B
5. A

Fig. 3-10

1. A
2. C
3. C
4. B
5. D

Fig. 3-11

1. C
2. A
3. A
4. B
5. D

Fig. 3-12

1. B
2. A
3. B
4. A
5. B

Fig. 3-13

1. D
2. B
3. B
4. B
5. A

Fig. 3-14

1. D
2. B
3. B
4. D
5. C

Fig. 3-15

1. B
2. D
3. C
4. C
5. C

Fig. 3-16

1. C
2. D
3. A
4. C
5. C

Fig. 3-17

1. D
2. A
3. A
4. C
5. B

Fig. 3-18

1. A
2. B
3. D
4. C
5. A

Fig. 3-19

1. A
2. C
3. A
4. B
5. B

Fig. 3-20

1. B
2. C
3. D
4. C
5. D

Fig. 3-21

1. C
2. B
3. D
4. A
5. C

Fig. 3-22

1. A
2. C
3. C
4. B
5. D

Fig. 3-23

1. A
2. C
3. C
4. B
5. A

Fig. 3-24

1. B
2. D
3. B
4. D
5. C

Fig. 3-25

1. A
2. B
3. B
4. C
5. A

Fig. 3-26

1. C
2. A
3. B
4. D
5. D

Fig. 3-27

1. D
2. C
3. B
4. C
5. C

Fig. 3-28

1. A
2. A
3. C
4. B
5. A

Fig. 3-29

1. A
2. D
3. C
4. B
5. D

Fig. 3-30

1. B
2. D
3. A
4. A
5. C

Fig. 3-31

1. A
2. D
3. B
4. B
5. C

Fig. 3-32

1. B
2. B
3. D
4. C
5. A

Fig. 3-33

1. C
2. A
3. D
4. D
5. C

Fig. 3-34

1. B
2. D
3. B
4. C
5. B

Fig. 3-35

1. C
2. B
3. B
4. D
5. D

Fig. 3-36

1. B
2. C
3. A
4. C
5. B

Fig. 3-37

1. D
2. B
3. A
4. B
5. D

Fig. 3-38

1. A
2. C
3. C
4. B
5. D

Fig. 3-39

1. B
2. D
3. B
4. C
5. A

Fig. 3-40

1. C
2. D
3. B
4. A
5. C

Fig. 3-41

1. D
2. C
3. D
4. A
5. D

Fig. 3-42

1. C
2. B
3. D
4. A
5. C

Fig. 3-43

1. B
2. A
3. C
4. A
5. B

Fig. 3-44

1. C
2. C
3. A
4. D
5. A

Fig. 3-45

1. B
2. C
3. A
4. B
5. A

Fig. 3-46

1. C
2. A
3. D
4. C
5. C

Fig. 3-47

1. B
2. A
3. D
4. B
5. D

Fig. 3-48

1. C
2. B
3. C
4. C
5. D

Fig. 3-49

1. D
2. B
3. D
4. C
5. B

Fig. 3-50

1. C
2. C
3. A
4. C
5. C

Fig. 3-51

1. B
2. A
3. B
4. B
5. B

Fig. 3-52

1. D
2. C
3. C
4. A
5. D

Fig. 3-53

1. B
2. B
3. A
4. B
5. A

Fig. 3-54

1. C
2. C
3. D
4. A
5. B

Fig. 3-55

1. B
2. A
3. A
4. B
5. C

Fig. 3-56

1. B
2. C
3. C
4. A
5. B

Chapter 4

Head

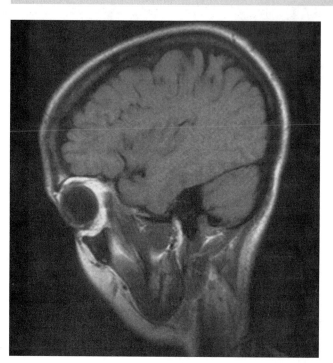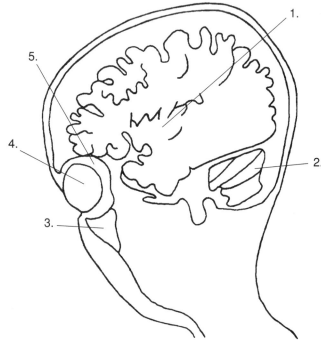

Figure 4-1

1. What number illustrates the insula?
   - _____ A. 5
   - _____ B. 2
   - _____ C. 1
   - _____ D. 3

2. Which of the following is illustrated by 3?
   - _____ A. Maxillary sinus
   - _____ B. Sphenoid sinus
   - _____ C. Frontal sinus
   - _____ D. Mastoid air cells

3. What number illustrates the cerebellum?
   - _____ A. 3
   - _____ B. 5
   - _____ C. 2
   - _____ D. 1

4. Which of the following is illustrated by 5?
   - _____ A. Sulci
   - _____ B. Inferior rectus muscle
   - _____ C. Fat in bony orbit
   - _____ D. Maxillary sinus

5. Which of the following is illustrated by 4?
   - _____ A. Globe of the eye
   - _____ B. Superior rectus muscle
   - _____ C. Optic nerve
   - _____ D. Inferior rectus muscle

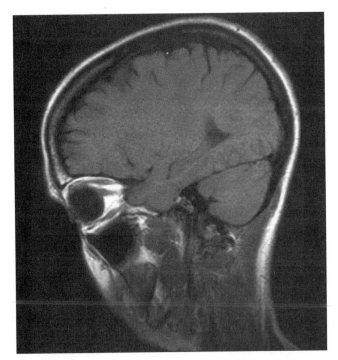
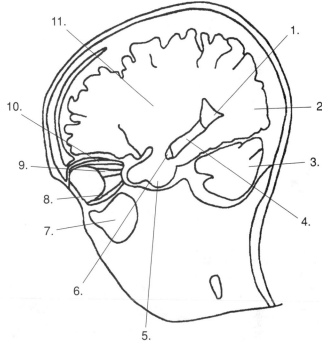

Figure 4-2

1. Which of the following is illustrated by 4?
   _____ A. Posterior horn of the lateral ventricle
   _____ B. Body of caudate nucleus
   _____ C. Thalamus
   _____ D. Inferior horn of the lateral ventricle

2. What number illustrates the inferior horn of the lateral ventricle?
   _____ A. 6
   _____ B. 4
   _____ C. 1
   _____ D. 7

3. Which of the following is illustrated by 1?
   _____ A. Body of caudate nucleus
   _____ B. Insula
   _____ C. Posterior horn of the lateral ventricle
   _____ D. Inferior horn of the lateral ventricle

4. Which of the following is illustrated by 9?
   _____ A. Superior rectus muscle
   _____ B. Inferior rectus muscle
   _____ C. Optic nerve
   _____ D. Fat in bony orbit

5. What number illustrates the superior rectus muscle?
   _____ A. 9
   _____ B. 6
   _____ C. 8
   _____ D. 10

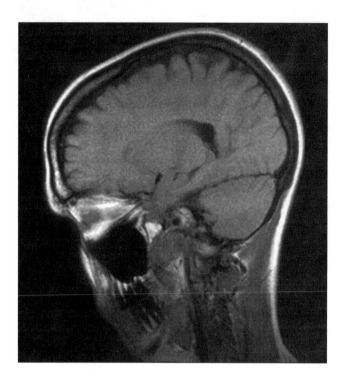
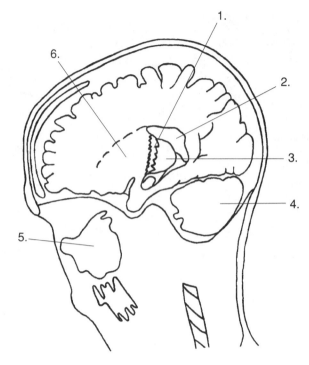

Figure 4-3

1. Which of the following is illustrated by 6?

    _____ A. Thalamus

    _____ B. Lenticular nuclei

    _____ C. Insula

    _____ D. Occipital lobe

2. What number illustrates the internal capsule?

    _____ A. 3

    _____ B. 2

    _____ C. 6

    _____ D. 1

3. Which of the following is illustrated by 5?

    _____ A. Maxillary sinus

    _____ B. Frontal sinus

    _____ C. Fat in bony orbit

    _____ D. Insula

4. What number illustrates the cerebellum?

    _____ A. 3

    _____ B. 4

    _____ C. 5

    _____ D. 2

5. Which of the following is illustrated by 3?

    _____ A. Lenticular nuclei

    _____ B. Body of caudate nucleus

    _____ C. Thalamus

    _____ D. Internal capsule

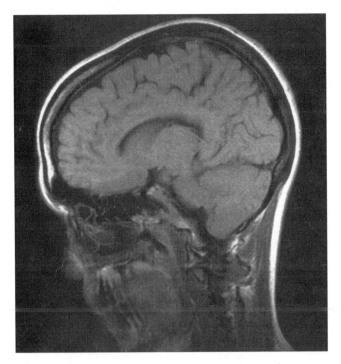 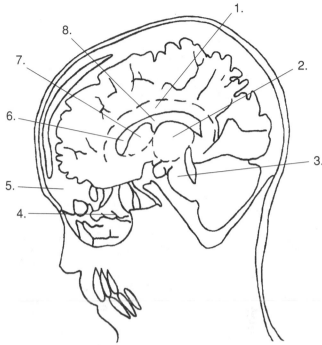

Figure 4-4

1. Which of the following is illustrated by 8?

    _____ A. Thalamus

    _____ B. Body of lateral ventricle

    _____ C. Head of caudate

    _____ D. Corpus callosum

2. What number illustrates the anterior horn of the lateral ventricle?

    _____ A. 8

    _____ B. 2

    _____ C. 7

    _____ D. 6

3. Which of the following is illustrated by 5?

    _____ A. Maxillary sinus

    _____ B. Ethmoid sinus

    _____ C. Frontal sinus

    _____ D. Fat in bony orbit

4. What number illustrates the corpus callosum?

    _____ A. 1

    _____ B. 8

    _____ C. 7

    _____ D. 6

5. Which of the following is illustrated by 7?

    _____ A. Thalamus

    _____ B. Head of caudate

    _____ C. Intermediate mass

    _____ D. Body of lateral ventricle

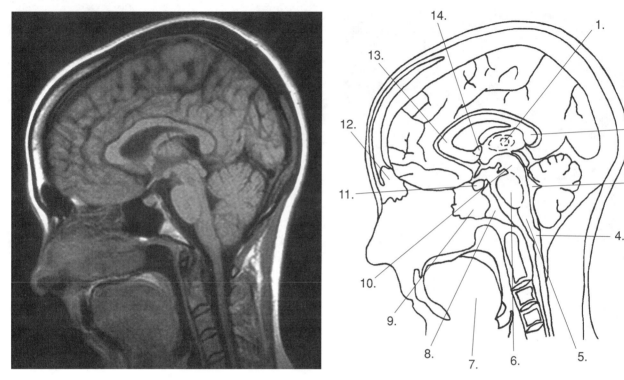

Figure 4-5

1. What number illustrates the splenium of corpus callosum?

    _____ A. 1

    _____ B. 13

    _____ C. 2

    _____ D. 14

2. Which of the following is illustrated by 10?

    _____ A. Pons

    _____ B. Cerebral peduncles

    _____ C. Pituitary gland

    _____ D. Quadrigeminal plate

3. What number illustrates the pituitary gland?

    _____ A. 11

    _____ B. 3

    _____ C. 10

    _____ D. 6

4. Which of the following is illustrated by 1?

    _____ A. Head of caudate nucleus

    _____ B. Anterior commissure

    _____ C. Intermediate mass

    _____ D. Splenium of corpus callosum

5. Which of the following is illustrated by 14?

    _____ A. Intermediate mass

    _____ B. Pituitary gland

    _____ C. Genu of corpus callosum

    _____ D. Anterior commissure

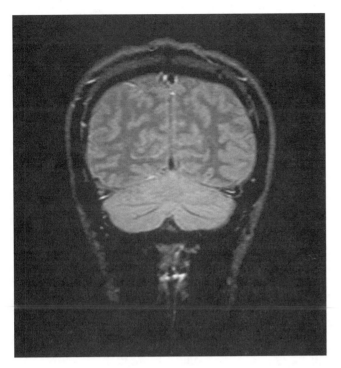
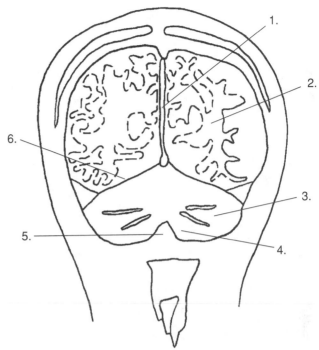

Figure 4-6

1. Which of the following is demonstrated by 2?

_____ A. Occipital lobe

_____ B. Parietal lobe

_____ C. Cerebellar hemisphere

_____ D. Temporal lobe

2. What number illustrates the cisterna magna?

_____ A. 6

_____ B. 1

_____ C. 5

_____ D. 4

3. Which of the following is demonstrated by 1?

_____ A. Straight sinus

_____ B. Falx cerebri

_____ C. Superior sagittal sinus

_____ D. Falx cerebelli

4. What number illustrates the cerebellar tonsil?

_____ A. 3

_____ B. 5

_____ C. 6

_____ D. 4

5. Which of the following is demonstrated by 3?

_____ A. Cerebellar hemisphere

_____ B. Cerebellar tonsil

_____ C. Occipital lobe

_____ D. Temporal lobe

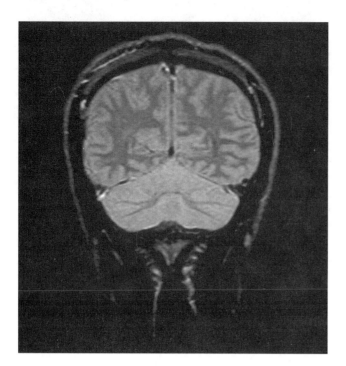
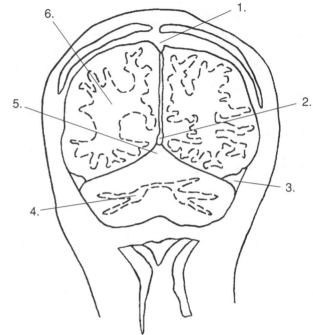

Figure 4-7

1. What number illustrates the occipital lobe white matter?
   _____ A. 3
   _____ B. 1
   _____ C. 6
   _____ D. 4

2. Which of the following is illustrated by 1?
   _____ A. Straight sinus
   _____ B. Superior sagittal sinus
   _____ C. Transverse sinus
   _____ D. Cisterna magna

3. What number illustrates the superior cerebellar vermis?
   _____ A. 5
   _____ B. 1
   _____ C. 3
   _____ D. 2

4. Which of the following is illustrated by 2?
   _____ A. Straight sinus
   _____ B. Superior sagittal sinus
   _____ C. Transverse sinus
   _____ D. Cisterna magna

5. Which of the following is illustrated by 3?
   _____ A. Superior sagittal sinus
   _____ B. Cisterna magna
   _____ C. Straight sinus
   _____ D. Transverse sinus

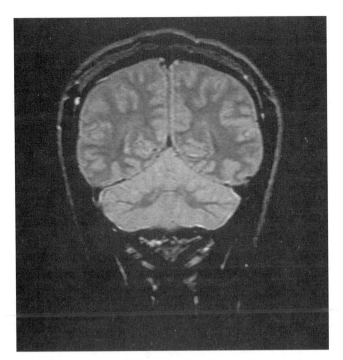
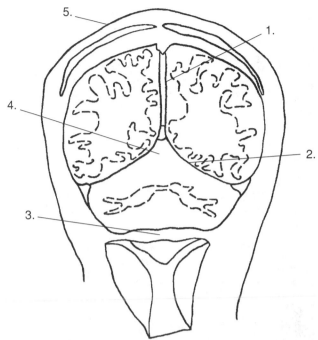

Figure 4-8

1. Which of the following is illustrated by 5?

    _____ A. Parietal bone

    _____ B. Scalp

    _____ C. Tentorium cerebelli

    _____ D. Occipital bone

2. What number illustrates the superior cerebellar vermis?

    _____ A. 4

    _____ B. 2

    _____ C. 3

    _____ D. 1

3. Which of the following is illustrated by 1?

    _____ A. Straight sinus

    _____ B. Superior sagittal sinus

    _____ C. Falx cerebri

    _____ D. Falx cerebelli

4. What number illustrates the tentorium cerebelli?

    _____ A. 3

    _____ B. 1

    _____ C. 4

    _____ D. 2

5. Which of the following is illustrated by 3?

    _____ A. Scalp

    _____ B. Occipital bone

    _____ C. Posterior medulla oblongata

    _____ D. Cisterna magna

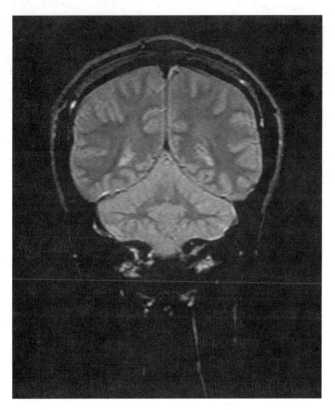 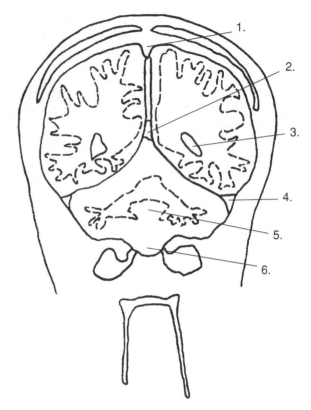

Figure 4-9

1. What number illustrates the fourth ventricle?

    _____ A. 3

    _____ B. 1

    _____ C. 5

    _____ D. 6

2. Which of the following is illustrated by 1?

    _____ A. Straight sinus

    _____ B. Superior sagittal sinus

    _____ C. Cisterna magna

    _____ D. Fourth ventricle

3. Which of the following is illustrated by 4?

    _____ A. Superior sagittal sinus

    _____ B. Transverse sinus

    _____ C. Straight sinus

    _____ D. Confluens of sinuses

4. What number illustrates the straight sinus?

    _____ A. 4

    _____ B. 5

    _____ C. 1

    _____ D. 2

5. Which of the following is illustrated by 3?

    _____ A. Posterior horn of the lateral ventricle

    _____ B. Anterior horn of lateral ventricle

    _____ C. Cisterna magna

    _____ D. Fourth ventricle

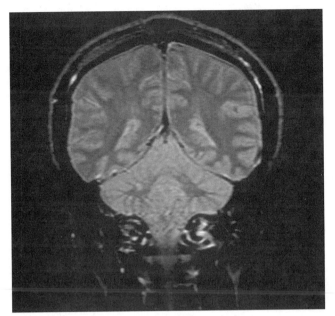 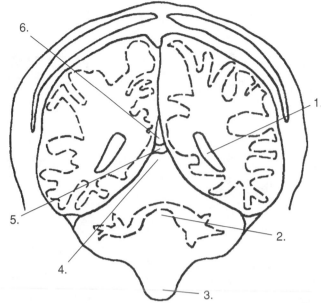

Figure 4-10

1. What number illustrates the fourth ventricle?

    _____ A. 6

    _____ B. 1

    _____ C. 5

    _____ D. 2

2. Which of the following is illustrated by 6?

    _____ A. Superior sagittal sinus

    _____ B. Straight sinus

    _____ C. Transverse sinus

    _____ D. Fourth ventricle

3. What number illustrates the superior cerebellar vermis?

    _____ A. 4

    _____ B. 6

    _____ C. 1

    _____ D. 5

4. Which of the following is illustrated by 5?

    _____ A. Superior cerebellar vermis

    _____ B. Straight sinus

    _____ C. Superior cistern

    _____ D. Fourth ventricle

5. Which of the following is illustrated by 1?

    _____ A. Posterior horn of the lateral ventricle

    _____ B. Transverse sinus

    _____ C. Straight sinus

    _____ D. Fourth ventricle

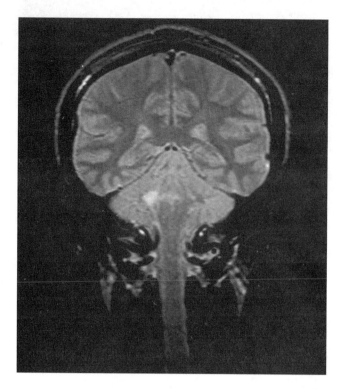
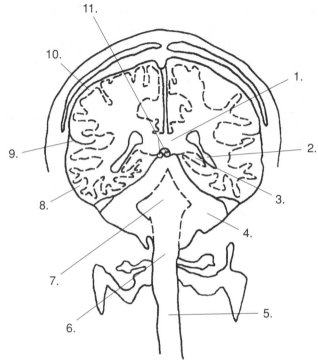

Figure 4-11

1. What number illustrates the posterior pons?

    _____ A. 11

    _____ B. 6

    _____ C. 5

    _____ D. 7

2. Which of the following is illustrated by 1?

    _____ A. Splenium of corpus callosum

    _____ B. Hippocampal formation

    _____ C. Anterior commissure

    _____ D. Genu of corpus callosum

3. Which of the following is illustrated by 9?

    _____ A. Sylvian fissure

    _____ B. Collateral trigone of the lateral ventricle

    _____ C. Splenium of corpus callosum

    _____ D. Superior cistern

4. What number illustrates the hippocampal formation?

    _____ A. 2

    _____ B. 1

    _____ C. 3

    _____ D. 11

5. What number illustrates the collateral trigone of the lateral ventricle?

    _____ A. 2

    _____ B. 9

    _____ C. 3

    _____ D. 1

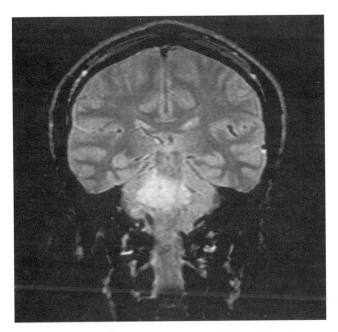 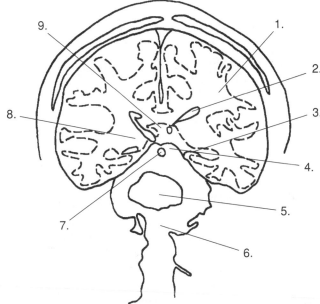

Figure 4-12

1. Which of the following is illustrated by 4?

    _____ A. Quadrigeminal plate

    _____ B. Cerebral aqueduct

    _____ C. Medulla oblongata

    _____ D. Corona radiata

2. What number illustrates the pineal gland?

    _____ A. 7

    _____ B. 9

    _____ C. 3

    _____ D. 2

3. What number illustrates the cerebral aqueduct?

    _____ A. 5

    _____ B. 4

    _____ C. 7

    _____ D. 9

4. Which of the following is illustrated by 1?

    _____ A. Quadrigeminal plate

    _____ B. Corona radiata

    _____ C. Thalamus

    _____ D. Body of lateral ventricle

5. What number illustrates the inferior horn of the lateral ventricle?

    _____ A. 7

    _____ B. 9

    _____ C. 2

    _____ D. 3

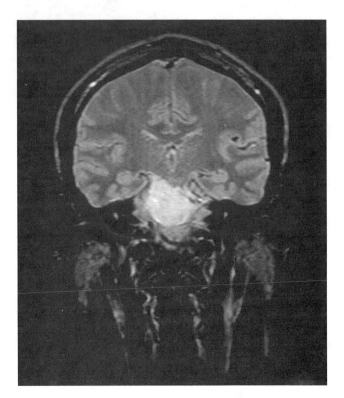
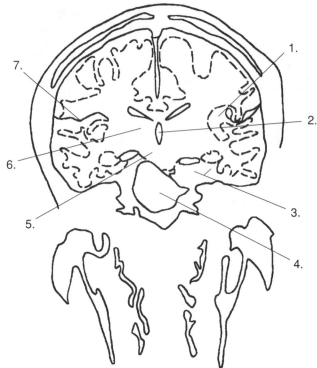

Figure 4-13

1. Which of the following is illustrated by 3?

_____ A. Hippocampal formation

_____ B. Sylvian fissure

_____ C. Head of caudate nucleus

_____ D. Quadrigeminal plate

2. What number illustrates the insula?

_____ A. 6

_____ B. 5

_____ C. 7

_____ D. 1

3. Which of the following is illustrated by 2?

_____ A. Pineal gland

_____ B. Third ventricle

_____ C. Pons

_____ D. Cerebral aqueduct

4. Which of the following is illustrated by 6?

_____ A. Sylvian fissure

_____ B. Thalamus

_____ C. Insula

_____ D. Lenticular nuclei

5. What number illustrates the cerebral peduncles?

_____ A. 3

_____ B. 6

_____ C. 1

_____ D. 5

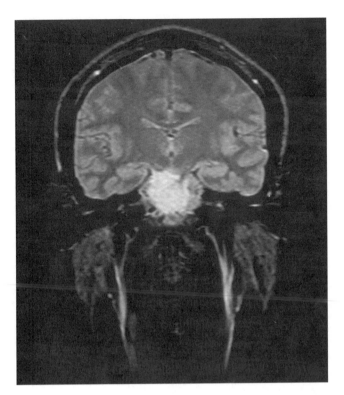
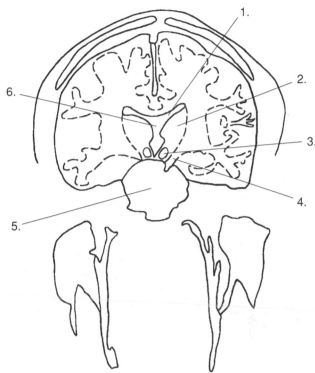

Figure 4-14

1. What number illustrates the thalamus?

    _____ A. 1

    _____ B. 3

    _____ C. 6

    _____ D. 2

2. Which of the following is illustrated by 4?

    _____ A. Red nucleus

    _____ B. Substantia nigra

    _____ C. Hippocampal formation

    _____ D. Quadrigeminal plate

3. What number illustrates the red nucleus?

    _____ A. 3

    _____ B. 6

    _____ C. 4

    _____ D. 1

4. Which of the following is illustrated by 1?

    _____ A. Insula

    _____ B. Thalamus

    _____ C. Third ventricle

    _____ D. Body of lateral ventricle

5. Which of the following is illustrated by 6?

    _____ A. Fourth ventricle

    _____ B. Collateral trigone of lateral ventricle

    _____ C. Third ventricle

    _____ D. Body of lateral ventricle

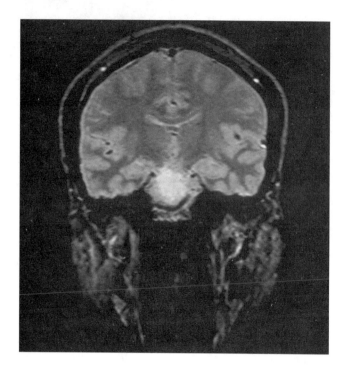
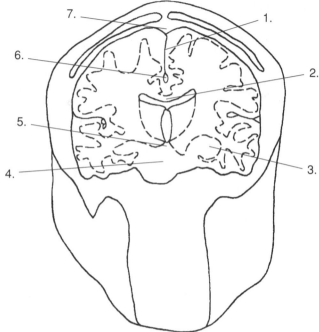

Figure 4-15

1. What number illustrates the body of corpus callosum?

    _____ A. 2

    _____ B. 1

    _____ C. 3

    _____ D. 4

2. Which of the following is illustrated by 4?

    _____ A. Hippocampal formation

    _____ B. Quadrigeminal plate

    _____ C. Midbrain

    _____ D. Pons

3. Which of the following is illustrated by 7?

    _____ A. Superior sagittal sinus

    _____ B. Falx cerebri

    _____ C. Cisterna magna

    _____ D. Inferior sagittal sinus

4. What number illustrates the midbrain?

    _____ A. 2

    _____ B. 1

    _____ C. 5

    _____ D. 4

5. Which of the following is illustrated by 6?

    _____ A. Superior sagittal sinus

    _____ B. Inferior sagittal sinus

    _____ C. Straight sinus

    _____ D. Superior cistern

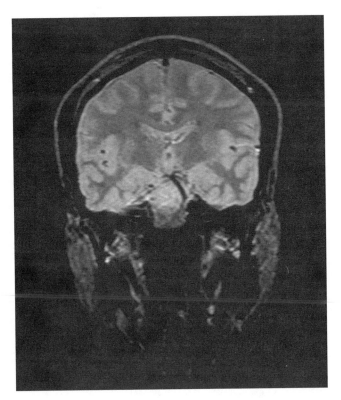
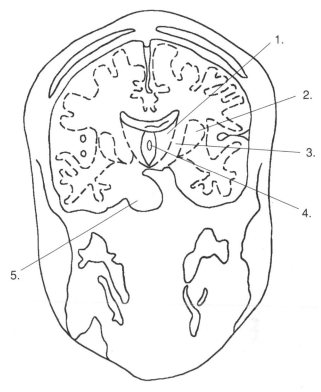

Figure 4-16

1. Which of the following is illustrated by 3?
   _____ A. Thalamus
   _____ B. Body of corpus callosum
   _____ C. Internal capsule
   _____ D. Lenticular nuclei

2. Which of the following is illustrated by 5?
   _____ A. Globus pallidus
   _____ B. Hippocampal formation
   _____ C. Midbrain
   _____ D. Anterior pons

3. Which of the following is illustrated by 2?
   _____ A. Lenticular nuclei
   _____ B. Body of corpus callosum
   _____ C. Internal capsule
   _____ D. Thalamus

4. Which of the following is illustrated by 4?
   _____ A. Superior cistern
   _____ B. Fourth ventricle
   _____ C. Cerebral aqueduct
   _____ D. Third ventricle

5. What number illustrates the thalamus?
   _____ A. 3
   _____ B. 1
   _____ C. 2
   _____ D. 4

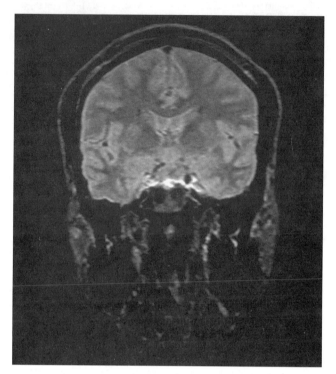

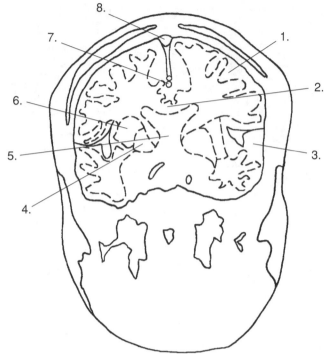

Figure 4-17

1. What number illustrates the thalamus?

    _____ A. 6

    _____ B. 4

    _____ C. 5

    _____ D. 7

2. Which of the following is illustrated by 1?

    _____ A. Superior sagittal sinus

    _____ B. Parietal lobe

    _____ C. Temporal lobe

    _____ D. Occipital lobe

3. Which of the following is illustrated by 6?

    _____ A. Body of corpus callosum

    _____ B. Globus pallidus

    _____ C. Thalamus

    _____ D. Putamen

4. What number illustrates the globus pallidus?

    _____ A. 4

    _____ B. 2

    _____ C. 6

    _____ D. 5

5. Which of the following is illustrated by 3?

    _____ A. Occipital lobe

    _____ B. Parietal lobe

    _____ C. Temporal lobe

    _____ D. Globus pallidus

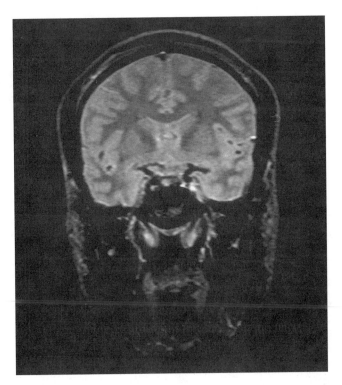
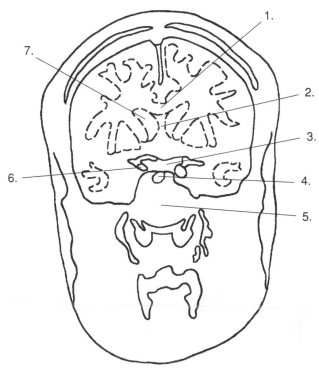

Figure 4-18

1. What number illustrates the anterior horn of the lateral ventricle?

   _____ A. 7

   _____ B. 2

   _____ C. 1

   _____ D. 5

2. Which of the following is illustrated by 7?

   _____ A. Anterior horn of the lateral ventricle

   _____ B. Body of corpus callosum

   _____ C. Head of the caudate nucleus

   _____ D. Thalamus

3. What number illustrates the optic chiasma?

   _____ A. 4

   _____ B. 6

   _____ C. 5

   _____ D. 3

4. Which of the following is illustrated by 4?

   _____ A. Pituitary gland

   _____ B. Sphenoid sinus

   _____ C. Third ventricle

   _____ D. Optic chiasma

5. Which of the following is illustrated by 6?

   _____ A. Optic nerve

   _____ B. Pituitary gland

   _____ C. Internal carotid artery

   _____ D. Vertebral artery

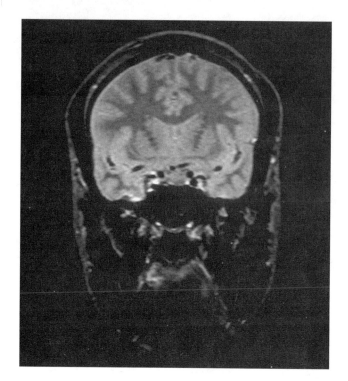 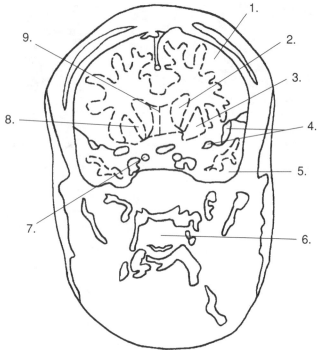

Figure 4-19

1. Which of the following is illustrated by 8?

    _____ A. Internal capsule

    _____ B. Head of caudate nucleus

    _____ C. Globus pallidus

    _____ D. Septum pellucidum

2. Which of the following is illustrated by 3?

    _____ A. Head of caudate nucleus

    _____ B. Septum pellucidum

    _____ C. Lenticular nuclei

    _____ D. Internal carotid artery

3. What number illustrates the septum pellucidum?

    _____ A. 8

    _____ B. 3

    _____ C. 2

    _____ D. 9

4. What number illustrates the internal carotid artery?

    _____ A. 7

    _____ B. 3

    _____ C. 2

    _____ D. 4

5. Which of the following is illustrated by 6?

    _____ A. Sphenoid sinus

    _____ B. Nasopharynx

    _____ C. Mastoid sinus

    _____ D. Oropharynx

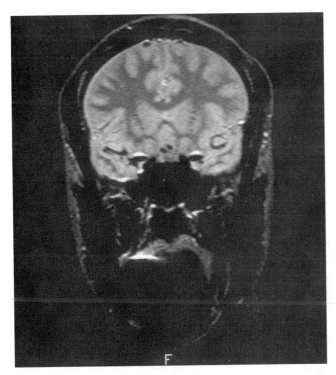 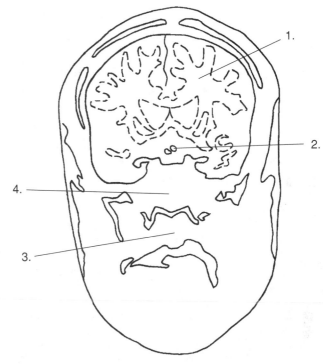

Figure 4-20

1. Which of the following is illustrated by 1?
   _____ A. White matter of parietal lobe
   _____ B. White matter of frontal lobe
   _____ C. Internal capsule
   _____ D. External capsule

2. Which of the following is illustrated by 2?
   _____ A. Straight sinus
   _____ B. Inferior sagittal sinus
   _____ C. Anterior cerebral artery
   _____ D. Middle cerebral artery

3. Which of the following is illustrated by 4?
   _____ A. Sphenoid sinus
   _____ B. Ethmoid air cells
   _____ C. Nasopharynx
   _____ D. Oropharynx

4. What number illustrates the nasopharynx?
   _____ A. 4
   _____ B. 3
   _____ C. 2
   _____ D. 1

5. Which of the following is illustrated by 3?
   _____ A. Sphenoid sinus
   _____ B. Nasopharynx
   _____ C. Oropharynx
   _____ D. Ethmoid air cells

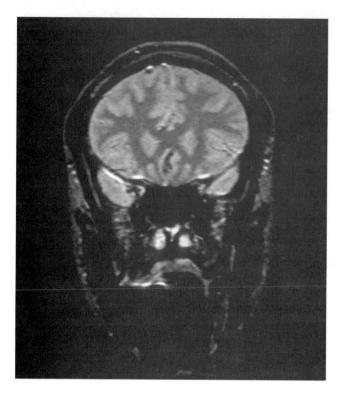
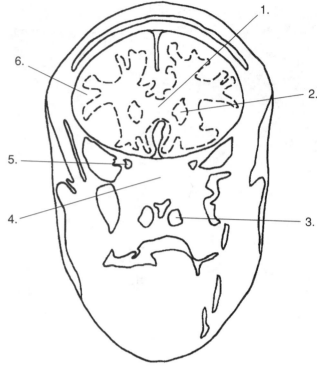

Figure 4-21

1. Which of the following is illustrated by 2?

    _____ A. Tail of caudate nucleus

    _____ B. Head of caudate nucleus

    _____ C. Lenticular nuclei

    _____ D. Thalamus

2. What number illustrates the inferior nasal concha?

    _____ A. 5

    _____ B. 2

    _____ C. 4

    _____ D. 3

3. Which of the following is illustrated by 4?

    _____ A. Sphenoid sinus

    _____ B. Nasopharynx

    _____ C. Maxillary sinus

    _____ D. Oropharynx

4. Which of the following is illustrated by 1?

    _____ A. Head of caudate nucleus

    _____ B. Genu of corpus callosum

    _____ C. Lenticular nuclei

    _____ D. Septum pellucidum

5. Which of the following is illustrated by 5?

    _____ A. Anterior cerebral artery

    _____ B. Internal carotid artery

    _____ C. Optic nerve

    _____ D. Posterior cerebral artery

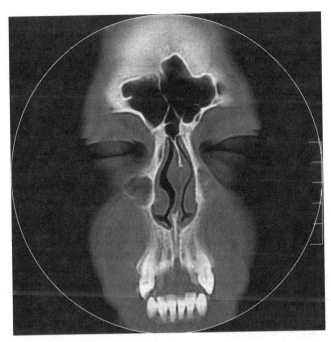
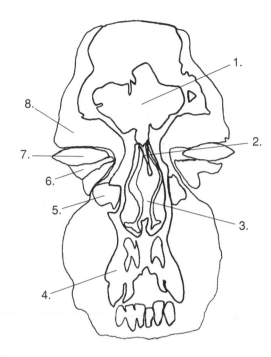

Figure 4-22

1. Which of the following is illustrated by 3?
    _____ A. Septal cartilage
    _____ B. Vomer
    _____ C. Perpendicular plate of ethmoid
    _____ D. Inferior concha

2. Which of the following is illustrated by 5?
    _____ A. Sphenoid sinus
    _____ B. Ethmoid sinus
    _____ C. Maxillary sinus
    _____ D. Frontal sinus

3. What number illustrates the perpendicular plate of ethmoid?
    _____ A. 3
    _____ B. 2
    _____ C. 4
    _____ D. 1

4. What number illustrates the frontal sinus?
    _____ A. 5
    _____ B. 3
    _____ C. 4
    _____ D. 1

5. What number illustrates the maxilla?
    _____ A. 4
    _____ B. 8
    _____ C. 2
    _____ D. 1

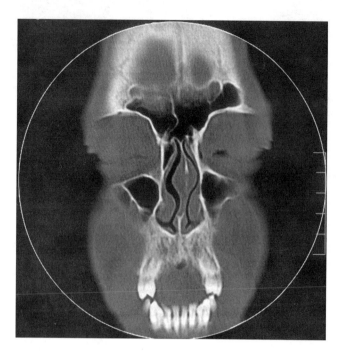
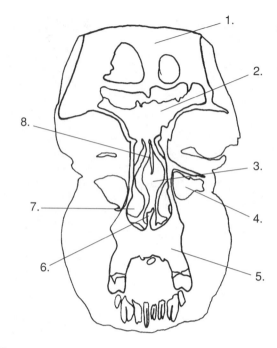

Figure 4-23

1. Which of the following is illustrated by 8?

    _____ A. Septal cartilage

    _____ B. Vomer

    _____ C. Perpendicular plate of ethmoid

    _____ D. Inferior concha

2. Which of the following is illustrated by 6?

    _____ A. Septal cartilage

    _____ B. Vomer

    _____ C. Perpendicular plate of ethmoid

    _____ D. Inferior concha

3. What number illustrates the inferior concha?

    _____ A. 7

    _____ B. 8

    _____ C. 6

    _____ D. 3

4. What number illustrates the septal cartilage?

    _____ A. 7

    _____ B. 8

    _____ C. 6

    _____ D. 3

5. Which of the following is illustrated by 5?

    _____ A. Palatine

    _____ B. Sphenoid

    _____ C. Maxilla

    _____ D. Ethmoid

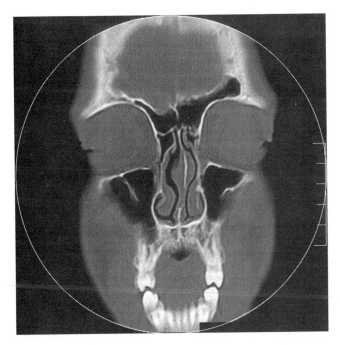
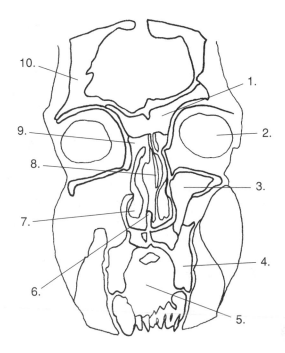

Figure 4-24

1. Which of the following is illustrated by 9?

    _____ A. Sphenoid sinus

    _____ B. Ethmoid sinus

    _____ C. Frontal sinus

    _____ D. Maxillary sinus

2. Which of the following is illustrated by 1?

    _____ A. Sphenoid sinus

    _____ B. Ethmoid sinus

    _____ C. Frontal sinus

    _____ D. Maxillary sinus

3. Which of the following is illustrated by 7?

    _____ A. Inferior concha

    _____ B. Middle concha

    _____ C. Vomer

    _____ D. Septal cartilage

4. What number illustrates the ethmoid sinus?

    _____ A. 1

    _____ B. 3

    _____ C. 9

    _____ D. 4

5. What number illustrates the maxillary sinus?

    _____ A. 1

    _____ B. 3

    _____ C. 9

    _____ D. 4

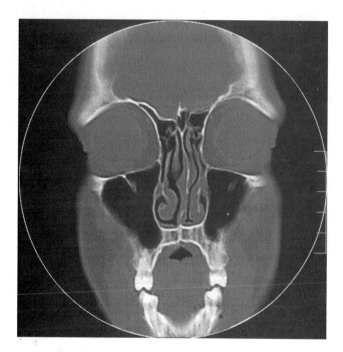
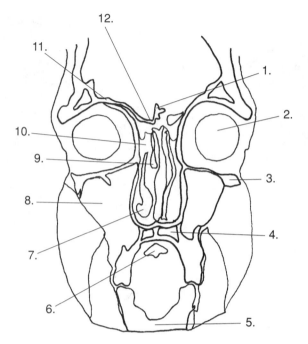

Figure 4-25

1. Which of the following is illustrated by 9?

    _____ A. Superior concha

    _____ B. Middle concha

    _____ C. Inferior concha

    _____ D. Ethmoid sinus

2. Which of the following is illustrated by 1?

    _____ A. Crista galli

    _____ B. Perpendicular plate of ethmoid

    _____ C. Vomer

    _____ D. Cribriform plate

3. Which of the following is illustrated by 12?

    _____ A. Crista galli

    _____ B. Vomer

    _____ C. Orbital plate of frontal bone

    _____ D. Cribriform plate

4. What number illustrates the inferior concha?

    _____ A. 10

    _____ B. 9

    _____ C. 7

    _____ D. 4

5. What number illustrates the ethmoid sinus?

    _____ A. 9

    _____ B. 10

    _____ C. 8

    _____ D. 4

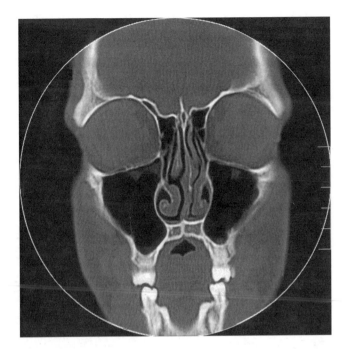
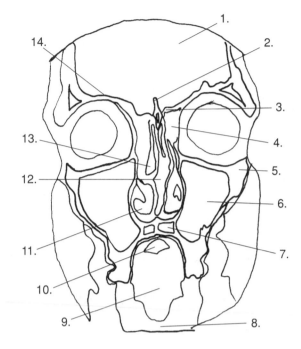

Figure 4-26

1. Which of the following is illustrated by 6?

    _____ A. Sphenoid sinus

    _____ B. Ethmoid sinus

    _____ C. Maxillary sinus

    _____ D. Frontal sinus

2. Which of the following is illustrated by 13?

    _____ A. Superior concha

    _____ B. Middle concha

    _____ C. Inferior concha

    _____ D. Nasal septum

3. Which of the following is illustrated by 5?

    _____ A. Ethmoid

    _____ B. Zygoma

    _____ C. Maxilla

    _____ D. Palatine

4. What number illustrates the cribriform plate?

    _____ A. 2

    _____ B. 3

    _____ C. 14

    _____ D. 7

5. What number illustrates the crista galli?

    _____ A. 2

    _____ B. 3

    _____ C. 4

    _____ D. 14

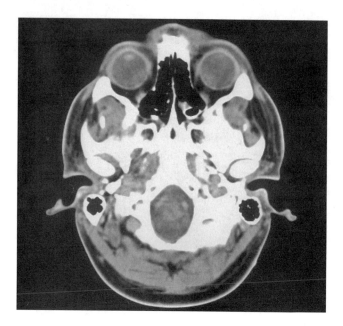
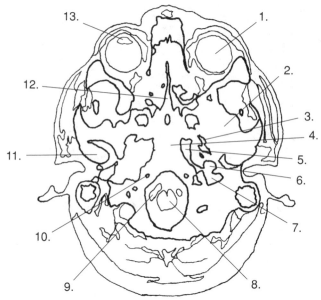

Figure 4-27

1. What number illustrates the foramen ovale?

    _____ A. 5

    _____ B. 3

    _____ C. 6

    _____ D. 7

2. Which of the following is illustrated by 7?

    _____ A. Internal jugular vein

    _____ B. Foramen lacerum

    _____ C. Foramen ovale

    _____ D. Internal carotid artery

3. What number illustrates the hypoglossal canal?

    _____ A. 6

    _____ B. 7

    _____ C. 10

    _____ D. 5

4. Which of the following illustrates the foramen lacerum?

    _____ A. 7

    _____ B. 6

    _____ C. 10

    _____ D. 5

5. Which of the following is illustrated by 6?

    _____ A. Foramen lacerum

    _____ B. Internal carotid artery

    _____ C. Foramen ovale

    _____ D. Internal jugular vein

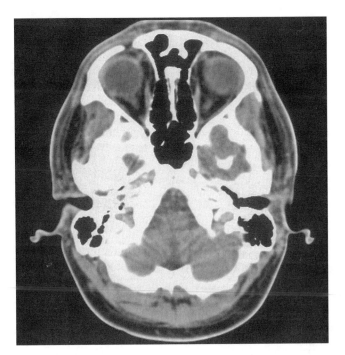 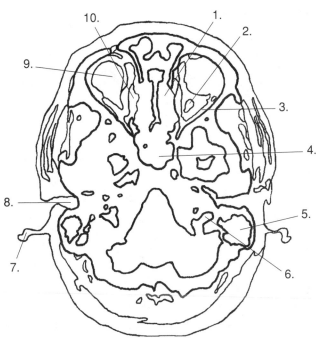

Figure 4-28

1. Which of the following is illustrated by 3?

    _____ A. Lateral rectus muscle

    _____ B. Medial rectus muscle

    _____ C. Optic nerve

    _____ D. Superior oblique muscle

2. What number illustrates the ethmoid air cells?

    _____ A. 4

    _____ B. 1

    _____ C. 8

    _____ D. 5

3. What number illustrates the medial rectus muscle?

    _____ A. 10

    _____ B. 3

    _____ C. 1

    _____ D. 2

4. Which of the following is illustrated by 5?

    _____ A. Temporal lobe

    _____ B. Sigmoid sinus

    _____ C. Internal carotid artery

    _____ D. Mastoid air cells

5. Which of the following is illustrated by 8?

    _____ A. Auricle

    _____ B. External acoustic meatus

    _____ C. Mandibular condyle

    _____ D. Hypoglossal canal

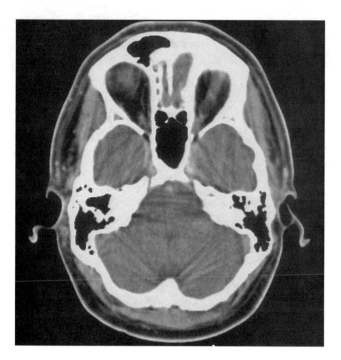

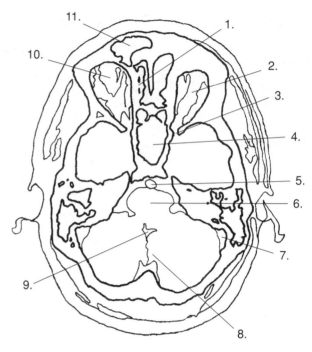

Figure 4-29

1. What number illustrates the cribriform plate of the ethmoid bone?

_____ A. 2

_____ B. 11

_____ C. 4

_____ D. 1

2. Which of the following is illustrated by 6?

_____ A. Pons

_____ B. Medulla oblongata

_____ C. Cerebellar vermis

_____ D. Cerebral peduncles

3. What number illustrates the basilar artery?

_____ A. 6

_____ B. 4

_____ C. 5

_____ D. 9

4. Which of the following is illustrated by 9?

_____ A. Fourth ventricle

_____ B. Basilar artery

_____ C. Superior cistern

_____ D. Inferior sagittal sinus

5. Which of the following is illustrated by 4?

_____ A. Ethmoid air cells

_____ B. Mastoid air cells

_____ C. Sigmoid sinus

_____ D. Sphenoid sinus

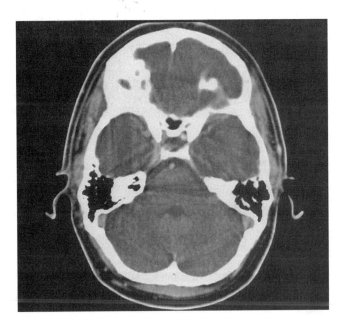
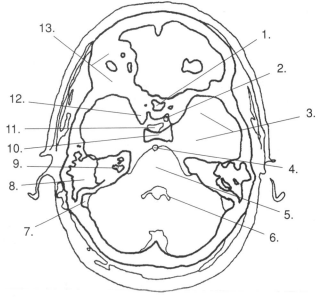

Figure 4-30

1. What number illustrates the dorsum sellae?

_____ A. 12

_____ B. 10

_____ C. 8

_____ D. 11

2. Which of the following is illustrated by 1?

_____ A. Optic chiasma

_____ B. Internal carotid artery

_____ C. Basilar artery

_____ D. Sphenoid sinus

3. What number illustrates the middle ear?

_____ A. 4

_____ B. 11

_____ C. 9

_____ D. 10

4. Which of the following is illustrated by 4?

_____ A. Basilar artery

_____ B. Interior carotid artery

_____ C. Middle ear

_____ D. Pituitary gland

5. Which of the following is illustrated by 11?

_____ A. Pituitary gland

_____ B. Dorsum sellae

_____ C. Optic chiasma

_____ D. Pineal gland

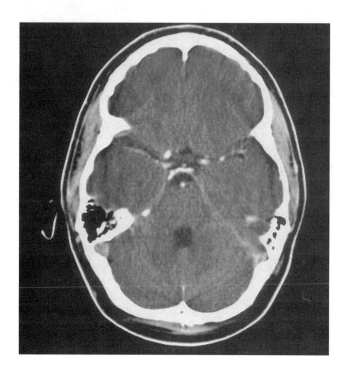

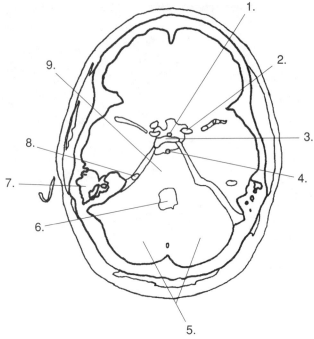

Figure 4-31

1. What number illustrates the infundibulum of the pituitary?

_____ A. 2

_____ B. 4

_____ C. 8

_____ D. 1

2. Which of the following is illustrated by 3?

_____ A. Basilar artery

_____ B. Middle cerebral artery

_____ C. Posterior cerebral artery

_____ D. Dorsum sellae

3. What number illustrates the mastoid air cells?

_____ A. 7

_____ B. 6

_____ C. 3

_____ D. 8

4. Which of the following is illustrated by 2?

_____ A. Infundibulum of the pituitary

_____ B. Middle cerebral artery

_____ C. Anterior cerebral artery

_____ D. Basilar artery

5. What number illustrates the posterior cerebral artery?

_____ A. 8

_____ B. 6

_____ C. 1

_____ D. 4

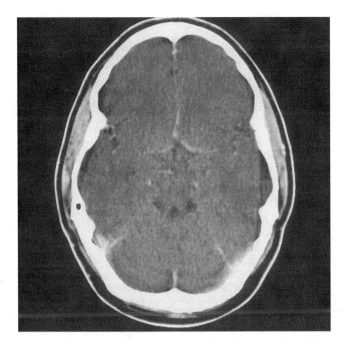 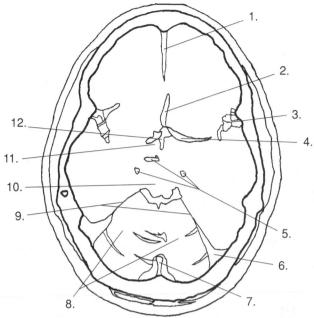

Figure 4-32

1. Which of the following is illustrated by 9?

    _____ A. Midbrain

    _____ B. Transverse sinus

    _____ C. Tentorium cerebelli

    _____ D. Cerebellum

2. What number illustrates the falx cerebri?

    _____ A. 7

    _____ B. 2

    _____ C. 5

    _____ D. 1

3. Which of the following illustrates the hypothalamus?

    _____ A. 12

    _____ B. 11

    _____ C. 8

    _____ D. 10

4. What number illustrates the Sylvian fissure?

    _____ A. 3

    _____ B. 1

    _____ C. 9

    _____ D. 12

5. Which of the following illustrate the falx cerebelli?

    _____ A. 1

    _____ B. 4

    _____ C. 7

    _____ D. 2

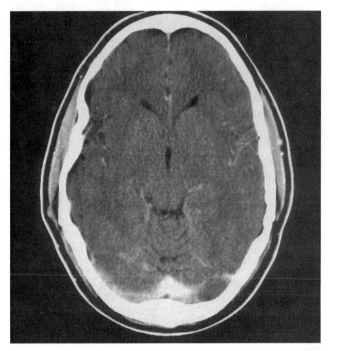
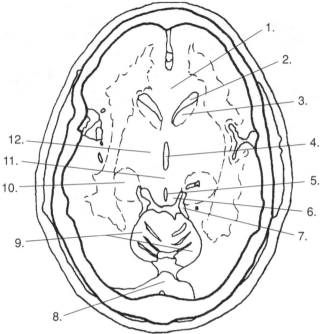

Figure 4-33

1. What number illustrates the quadrigeminal plate?

    _____ A. 6

    _____ B. 2

    _____ C. 7

    _____ D. 3

2. What number illustrates the cerebral peduncles?

    _____ A. 6

    _____ B. 11

    _____ C. 10

    _____ D. 12

3. Which of the following is illustrated by 4?

    _____ A. Fourth ventricle

    _____ B. Cerebral aqueduct

    _____ C. Third ventricle

    _____ D. Superior cistern

4. What number illustrates the hippocampal formation?

    _____ A. 3

    _____ B. 6

    _____ C. 11

    _____ D. 10

5. What number illustrates the posterior cerebral artery?

    _____ A. 7

    _____ B. 4

    _____ C. 8

    _____ D. 5

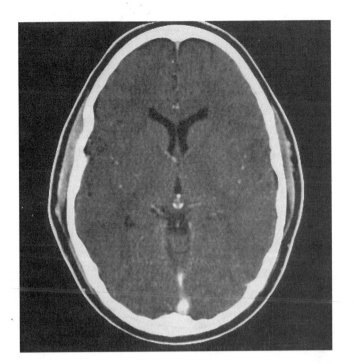
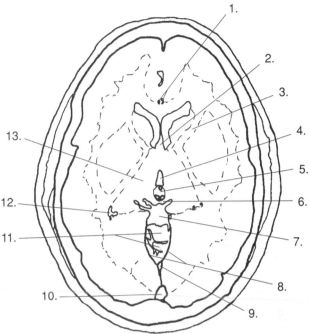

Figure 4-34

1. Which of the following is illustrated by 2?

    _____ A. Thalamus

    _____ B. Internal capsule

    _____ C. Fornix

    _____ D. Head of caudate

2. What number illustrates the quadrigeminal plate?

    _____ A. 6

    _____ B. 8

    _____ C. 7

    _____ D. 11

3. What number illustrates the inferior horn of the lateral ventricle?

    _____ A. 8

    _____ B. 5

    _____ C. 4

    _____ D. 12

4. Which of the following is illustrated by 9?

    _____ A. Confluence of sinuses

    _____ B. Straight sinus

    _____ C. Inferior sagittal sinus

    _____ D. Superior sagittal sinus

5. Which of the following is illustrated by 7?

    _____ A. Quadrigeminal plate

    _____ B. Superior cistern

    _____ C. Third ventricle

    _____ D. Straight sinus

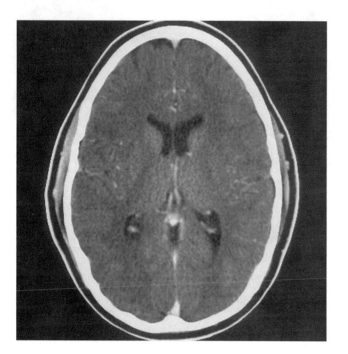

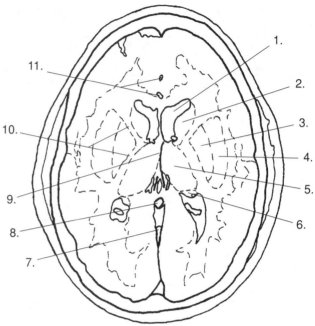

Figure 4-35

1. What number illustrates the thalamus nucleus?

     _____ A. 2

     _____ B. 4

     _____ C. 3

     _____ D. 5

2. Which of the following is illustrated by 8?

     _____ A. Posterior cerebral artery

     _____ B. Vein of Galen

     _____ C. Third ventricle

     _____ D. Anterior cerebral artery

3. What number illustrates the straight sinus?

     _____ A. 9

     _____ B. 6

     _____ C. 7

     _____ D. 1

4. Which of the following is illustrated by 4?

     _____ A. Head of caudate

     _____ B. Globus pallidus

     _____ C. Putamen

     _____ D. Tail of caudate

5. What number illustrates the globus pallidus nucleus?

     _____ A. 2

     _____ B. 4

     _____ C. 3

     _____ D. 6

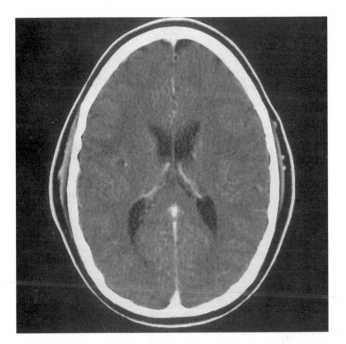 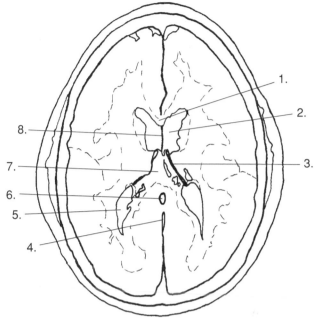

Figure 4-36

1. Which of the following is illustrated by 2?

   _____ A. Anterior horn of lateral ventricle

   _____ B. Body of lateral ventricle

   _____ C. Third ventricle

   _____ D. Superior cistern

2. What number illustrates the septum pellucidum?

   _____ A. 1

   _____ B. 4

   _____ C. 3

   _____ D. 8

3. Which of the following is illustrated by 7?

   _____ A. Pons

   _____ B. Splenium of corpus callosum

   _____ C. Hypothalamus

   _____ D. Thalamus

4. What number illustrates the posterior horn of the lateral ventricle?

   _____ A. 3

   _____ B. 6

   _____ C. 5

   _____ D. 2

5. Which of the following is illustrated by 3?

   _____ A. Posterior horn of the lateral ventricle

   _____ B. Choroid plexus

   _____ C. Splenium of corpus callosum

   _____ D. Septum pellucidum

ANSWERS

Fig. 4-1	Fig. 4-7	Fig. 4-13	Fig. 4-19
1. C	1. C	1. A	1. A
2. A	2. B	2. D	2. C
3. C	3. A	3. B	3. D
4. C	4. A	4. B	4. A
5. A	5. D	5. D	5. B

Fig. 4-2	Fig. 4-8	Fig. 4-14	Fig. 4-20
1. B	1. B	1. D	1. B
2. A	2. A	2. B	2. C
3. C	3. C	3. A	3. A
4. C	4. D	4. D	4. B
5. D	5. B	5. C	5. B

Fig. 4-3	Fig. 4-9	Fig. 4-15	Fig. 4-21
1. B	1. C	1. A	1. B
2. D	2. B	2. D	2. D
3. A	3. B	3. A	3. A
4. B	4. D	4. C	4. B
5. C	5. A	5. B	5. C

Fig. 4-4	Fig. 4-10	Fig. 4-16	Fig. 4-22
1. B	1. D	1. C	1. A
2. D	2. B	2. D	2. C
3. C	3. A	3. A	3. B
4. A	4. C	4. D	4. D
5. B	5. A	5. B	5. A

Fig. 4-5	Fig. 4-11	Fig. 4-17	Fig. 4-23
1. C	1. D	1. C	1. C
2. B	2. A	2. B	2. B
3. A	3. A	3. D	3. A
4. C	4. C	4. A	4. D
5. D	5. A	5. C	5. C

Fig. 4-6	Fig. 4-12	Fig. 4-18	Fig. 4-24
1. A	1. A	1. B	1. B
2. C	2. B	2. C	2. C
3. B	3. C	3. D	3. A
4. D	4. B	4. A	4. C
5. A	5. D	5. C	5. B

Fig. 4-25

1. B
2. A
3. D
4. C
5. B

Fig. 4-26

1. C
2. B
3. B
4. B
5. A

Fig. 4-27

1. B
2. A
3. C
4. D
5. B

Fig. 4-28

1. A
2. B
3. A
4. D
5. B

Fig. 4-29

1. D
2. A
3. C
4. A
5. D

Fig. 4-30

1. B
2. D
3. C
4. A
5. A

Fig. 4-31

1. D
2. D
3. A
4. B
5. A

Fig. 4-32

1. C
2. D
3. B
4. A
5. C

Fig. 4-33

1. A
2. B
3. C
4. D
5. A

Fig. 4-34

1. C
2. A
3. D
4. B
5. B

Fig. 4-35

1. D
2. B
3. C
4. C
5. C

Fig. 4-36

1. A
2. D
3. B
4. C
5. B

Chapter 5

Neck

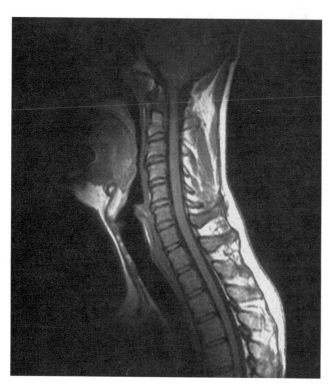
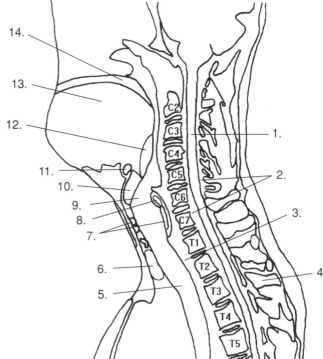

Figure 5-1

1. Which of the following is illustrated by 14?
   _____ A. Nasopharynx
   _____ B. Genioglossus muscle
   _____ C. Soft palate
   _____ D. Maxilla
2. Which of the following is illustrated by 7?
   _____ A. Thyroid cartilage
   _____ B. Cricoid cartilage
   _____ C. Arytenoid cartilage
   _____ D. Hyoid bone
3. What number illustrates the genioglossus muscle?
   _____ A. 13
   _____ B. 14
   _____ C. 12
   _____ D. 10

4. Which of the following is illustrated by 8?
   _____ A. Arytenoid cartilage
   _____ B. Cricoid cartilage
   _____ C. Thyroid cartilage
   _____ D. Hyoid bone
5. What number illustrates the isthmus of the thyroid gland?
   _____ A. 12
   _____ B. 7
   _____ C. 10
   _____ D. 6

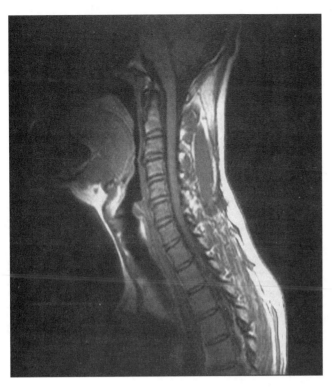
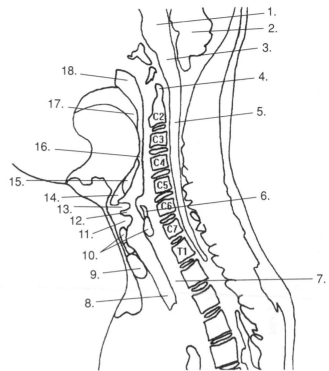

Figure 5-2

1. What number illustrates the vestibular fold?

    _____ A. 13

    _____ B. 6

    _____ C. 11

    _____ D. 18

2. Which of the following is illustrated by 15?

    _____ A. Oropharynx

    _____ B. Epiglottis

    _____ C. Cricoid cartilage

    _____ D. Arytenoid cartilage

3. Which of the following is illustrated by 12?

    _____ A. Vestibular fold

    _____ B. Uvula

    _____ C. Laryngeal pharynx

    _____ D. Glottic space

4. What number illustrates the thyroid cartilage?

    _____ A. 14

    _____ B. 15

    _____ C. 6

    _____ D. 10

5. What number illustrates the vocal fold?

    _____ A. 12

    _____ B. 13

    _____ C. 11

    _____ D. 15

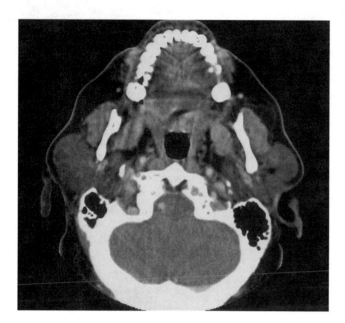 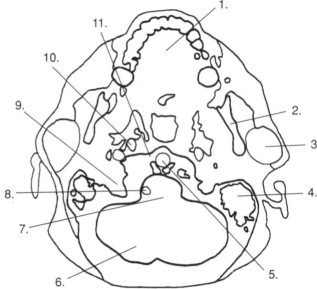

Figure 5-3

1. Which of the following is illustrated by 8?

    _____ A. Basilar artery

    _____ B. Posterior cerebral artery

    _____ C. Right vertebral artery

    _____ D. Right internal carotid artery

2. What number illustrates the sigmoid sinus?

    _____ A. 9

    _____ B. 10

    _____ C. 4

    _____ D. 5

3. Which of the following is illustrated by 3?

    _____ A. Submandibular gland

    _____ B. Mastoid air cells

    _____ C. Ramus of the mandible

    _____ D. Parotid gland

4. What number illustrates the internal carotid artery?

    _____ A. 5

    _____ B. 9

    _____ C. 10

    _____ D. 8

5. Which of the following is illustrated by 11?

    _____ A. Occipital bone

    _____ B. Dens

    _____ C. Styloid process

    _____ D. Anterior arch of the atlas

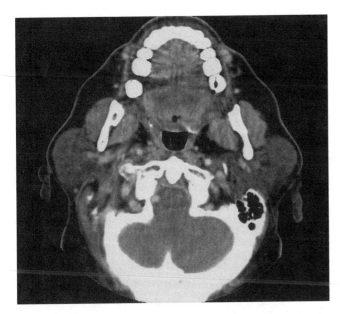 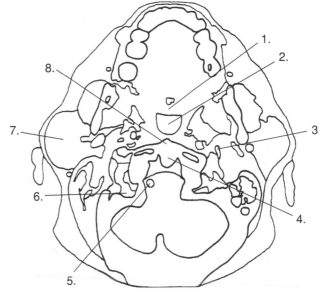

Figure 5-4

1. Which of the following is illustrated by 3?

    _____ A. Internal carotid artery

    _____ B. Styloid process

    _____ C. Internal jugular vein

    _____ D. Dens

2. What number illustrates the internal jugular vein?

    _____ A. 6

    _____ B. 5

    _____ C. 2

    _____ D. 3

3. Which of the following is illustrated by 7?

    _____ A. Retromandibular vein

    _____ B. Submandibular gland

    _____ C. Parotid gland

    _____ D. Sublingual gland

4. Which of the following is illustrated by 2?

    _____ A. Vallecula

    _____ B. Laryngeal pharynx

    _____ C. Uvula

    _____ D. Oropharynx

5. Which of the following is illustrated by 5?

    _____ A. Basilar artery

    _____ B. Internal jugular vein

    _____ C. Vertebral artery

    _____ D. Internal carotid artery

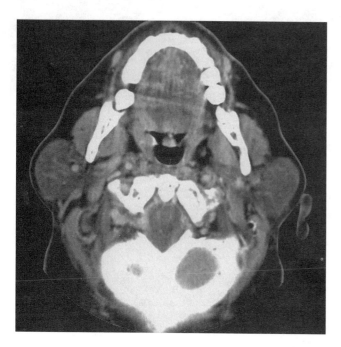
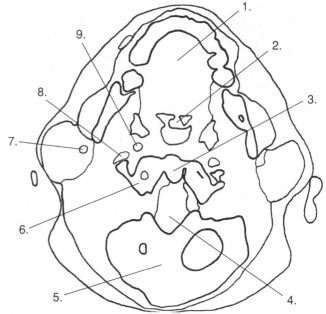

Figure 5-5

1. What number illustrates the uvula?

    _____ A. 6

    _____ B. 2

    _____ C. 4

    _____ D. 3

2. Which of the following is illustrated by 7?

    _____ A. Parotid gland

    _____ B. Right retromandibular vein

    _____ C. External jugular vein

    _____ D. Internal jugular vein

3. Which of the following is illustrated by 1?

    _____ A. Soft palate

    _____ B. Geniohyoid

    _____ C. Sternocleidomastoid

    _____ D. Genioglossus muscle

4. Which of the following is illustrated by 6?

    _____ A. Atlas

    _____ B. Dens

    _____ C. Axis

    _____ D. Occipital bone

5. Which of the following is illustrated by 4?

    _____ A. Spinal cord

    _____ B. Pons

    _____ C. Cerebral peduncles

    _____ D. Oropharynx

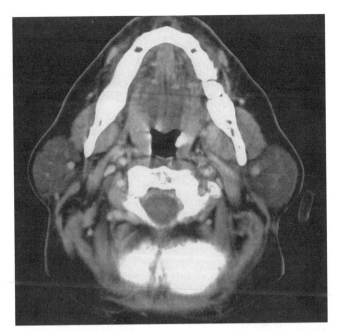
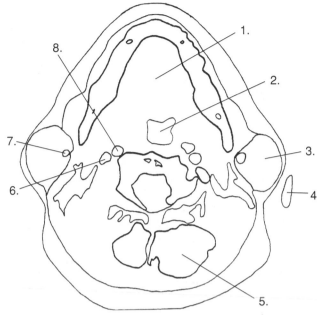

Figure 5-6

1. What number illustrates the oropharynx?

    _____ A. 8

    _____ B. 4

    _____ C. 2

    _____ D. 3

2. Which of the following is illustrated by 3?

    _____ A. Submandibular gland

    _____ B. Sublingual gland

    _____ C. Uvula

    _____ D. Parotid gland

3. Which number illustrates the internal jugular vein?

    _____ A. 6

    _____ B. 2

    _____ C. 8

    _____ D. 7

4. Which of the following is illustrated by 8?

    _____ A. Internal jugular vein

    _____ B. Retromandibular vein

    _____ C. Vertebral artery

    _____ D. Internal carotid artery

5. Which of the following is illustrated by 5?

    _____ A. Dens

    _____ B. Atlas

    _____ C. Occipital bone

    _____ D. Axis

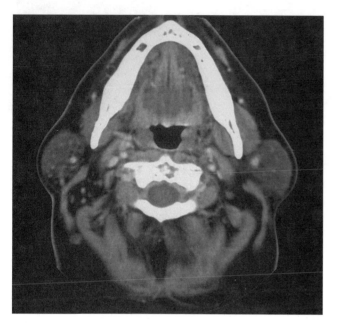
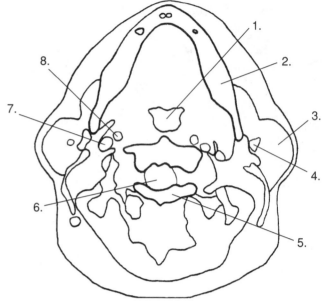

Figure 5-7

1. Which of the following is illustrated by 2?

    _____ A. Maxilla

    _____ B. Ramus of mandible

    _____ C. Body of the mandible

    _____ D. Coronoid notch

2. What number illustrates the lamina?

    _____ A. 5

    _____ B. 6

    _____ C. 2

    _____ D. 4

3. Which of the following is illustrated by 4?

    _____ A. Right retromandibular vein

    _____ B. Left retromandibular vein

    _____ C. Right external jugular vein

    _____ D. Left external jugular vein

4. What number illustrates the parotid gland?

    _____ A. 1

    _____ B. 6

    _____ C. 5

    _____ D. 3

5. What number illustrates the internal carotid artery?

    _____ A. 8

    _____ B. 4

    _____ C. 7

    _____ D. 6

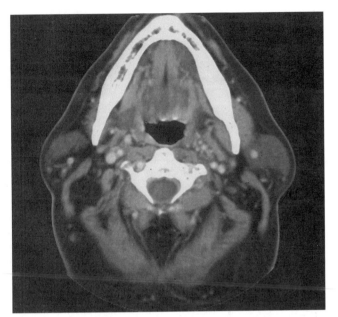 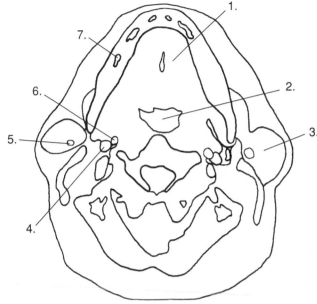

Figure 5-8

1. Which of the following is illustrated by 4?

    _____ A. Right internal carotid artery

    _____ B. Left internal carotid artery

    _____ C. Right internal jugular vein

    _____ D. Left internal jugular vein

2. Which of the following is illustrated by 1?

    _____ A. Soft palate

    _____ B. Geniohyoid muscle

    _____ C. Sternocleidomastoid muscle

    _____ D. Genioglossus muscle

3. Which of the following is illustrated by 7?

    _____ A. Body of mandible

    _____ B. Ramus of mandible

    _____ C. Maxilla

    _____ D. Coronoid notch

4. Which of the following is illustrated by 6?

    _____ A. Right internal jugular vein

    _____ B. Left internal jugular vein

    _____ C. Right internal carotid artery

    _____ D. Left retromandibular vein

5. Which of the following is illustrated by 5?

    _____ A. Right internal carotid artery

    _____ B. Left internal carotid artery

    _____ C. Left retromandibular vein

    _____ D. Right retromandibular vein

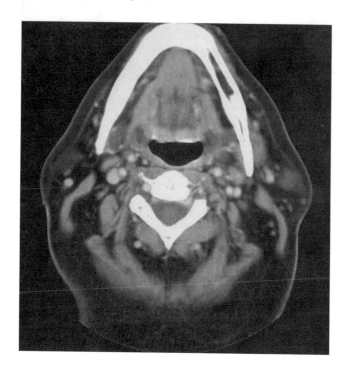

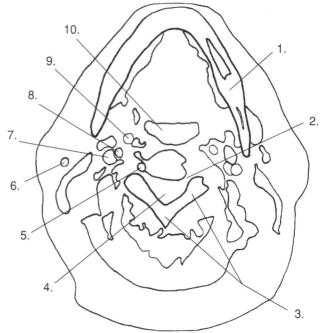

Figure 5-9

1. Which of the following illustrates the intervertebral foramen?

_____ A. 4

_____ B. 3

_____ C. 2

_____ D. 10

2. What number illustrates the internal jugular vein?

_____ A. 7

_____ B. 8

_____ C. 6

_____ D. 5

3. Which of the following is illustrated by 5?

_____ A. Internal carotid artery

_____ B. Internal jugular vein

_____ C. Retromandibular vein

_____ D. Vertebral artery

4. Which of the following is illustrated by 4?

_____ A. Lamina

_____ B. Spinal cord

_____ C. Intervertebral foramen

_____ D. Oropharynx

5. Which of the following is illustrated by 9?

_____ A. Internal carotid artery

_____ B. Internal jugular vein

_____ C. External carotid artery

_____ D. Vertebral artery

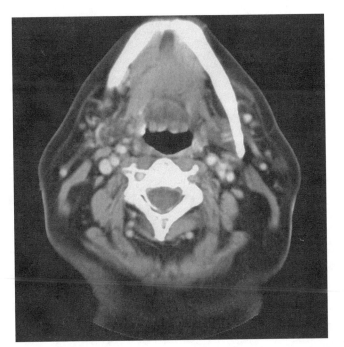 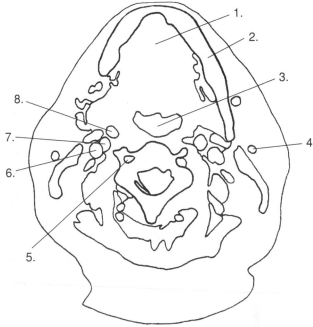

Figure 5-10

1. Which of the following is illustrated by 7?

    _____ A. Right internal carotid artery

    _____ B. Left internal carotid artery

    _____ C. Right external carotid artery

    _____ D. Left external carotid artery

2. What number illustrates the body of the mandible?

    _____ A. 3

    _____ B. 1

    _____ C. 2

    _____ D. 8

3. Which of the following is illustrated by 3?

    _____ A. Larynx

    _____ B. Laryngeal pharynx

    _____ C. Nasopharynx

    _____ D. Oropharynx

4. Which of the following is illustrated by 6?

    _____ A. Internal carotid artery

    _____ B. External carotid artery

    _____ C. Internal jugular vein

    _____ D. External jugular vein

5. Which of the following is illustrated by 4?

    _____ A. Retromandibular vein

    _____ B. Retromandibular artery

    _____ C. External jugular vein

    _____ D. External carotid artery

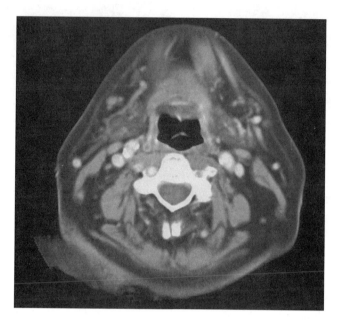
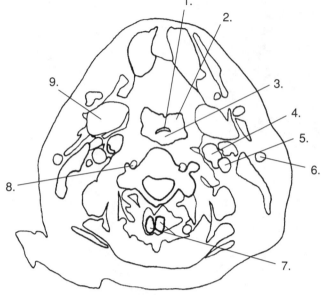

Figure 5-11

1. Which of the following is illustrated by 4?

    _____ A. Internal carotid artery

    _____ B. Common carotid artery

    _____ C. External jugular vein

    _____ D. Internal jugular vein

2. What number illustrates the epiglottis?

    _____ A. 7

    _____ B. 9

    _____ C. 1

    _____ D. 8

3. What number illustrates the internal jugular vein?

    _____ A. 5

    _____ B. 4

    _____ C. 6

    _____ D. 8

4. Which of the following is illustrated by 3?

    _____ A. Oropharynx

    _____ B. Laryngeal pharynx

    _____ C. Piriform sinus

    _____ D. Vallecula

5. Which of the following is illustrated by 2?

    _____ A. Vallecula

    _____ B. Glottic space

    _____ C. Laryngeal pharynx

    _____ D. Piriform sinus

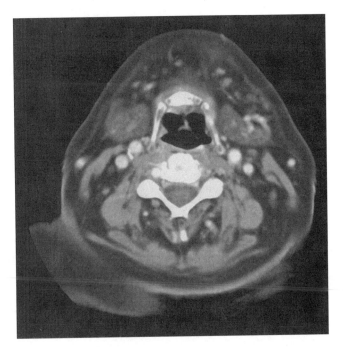

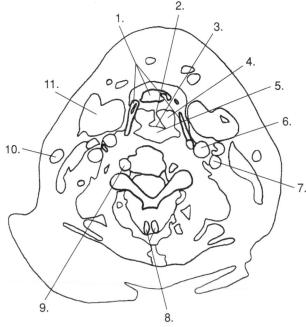

Figure 5-12

1. Which of the following is illustrated by 4?

_____ A. Glottic space

_____ B. Laryngeal pharynx

_____ C. Piriform sinus

_____ D. Vallecula

2. What number illustrates the laryngeal pharynx?

_____ A. 6

_____ B. 4

_____ C. 5

_____ D. 3

3. What number illustrates the median glossoepiglottic fold?

_____ A. 3

_____ B. 2

_____ C. 11

_____ D. 4

4. Which of the following is illustrated by 1?

_____ A. Hyoid bone

_____ B. Cricoid cartilage

_____ C. Body of mandible

_____ D. Thyroid cartilage

5. Which of the following is illustrated by 7?

_____ A. Common carotid artery

_____ B. Internal carotid artery

_____ C. Internal jugular vein

_____ D. External jugular vein

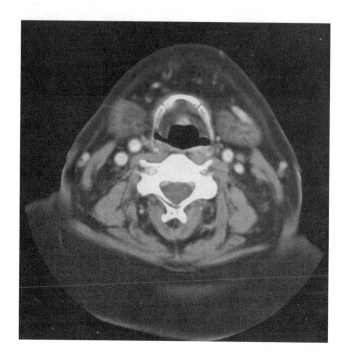
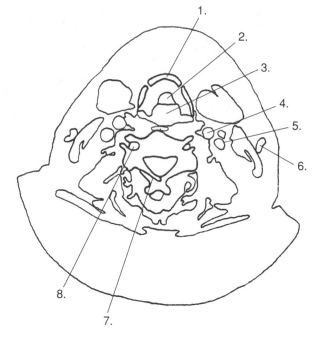

Figure 5-13

1. Which of the following is illustrated by 6?

    _____ A. External jugular vein

    _____ B. Internal jugular vein

    _____ C. Retromandibular vein

    _____ D. External carotid artery

2. What number illustrates the laryngeal vestibule?

    _____ A. 2

    _____ B. 3

    _____ C. 1

    _____ D. 8

3. Which of the following is illustrated by 2?

    _____ A. Arytenoid cartilage

    _____ B. Uvula

    _____ C. Epiglottis

    _____ D. Laryngeal vestibule

4. Which of the following is illustrated by 4?

    _____ A. Internal jugular vein

    _____ B. Internal carotid artery

    _____ C. Common carotid artery

    _____ D. External jugular vein

5. Which of the following is illustrated by 1?

    _____ A. Cricoid cartilage

    _____ B. Arytenoid cartilage

    _____ C. Thyroid cartilage

    _____ D. Hyoid bone

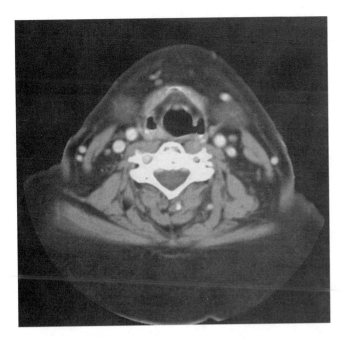 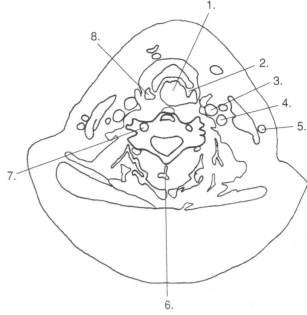

Figure 5-14

1. What number illustrates the vertebral artery?

    _____ A. 3

    _____ B. 4

    _____ C. 7

    _____ D. 5

2. Which of the following is illustrated by 1?

    _____ A. Piriform sinus

    _____ B. Laryngeal vestibule

    _____ C. Vallecula

    _____ D. Laryngeal pharynx

3. Which of the following is illustrated by 8?

    _____ A. Vallecula

    _____ B. Laryngeal pharynx

    _____ C. Piriform sinus

    _____ D. Laryngeal vestibule

4. Which of the following is illustrated by 2?

    _____ A. Aryepiglottic fold

    _____ B. Vestibular fold

    _____ C. Vallecula

    _____ D. Vocal fold

5. What number illustrates the common carotid artery?

    _____ A. 7

    _____ B. 5

    _____ C. 4

    _____ D. 3

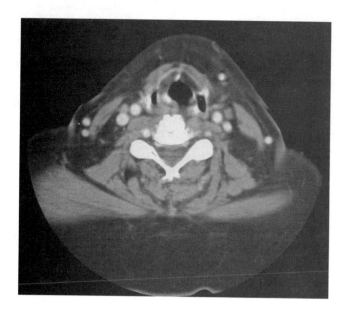
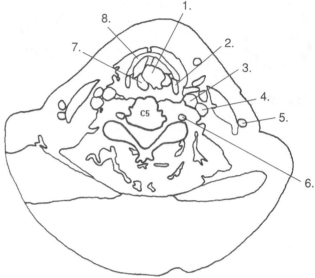

Figure 5-15

1. Which of the following is illustrated by 6?
   _____ A. Vertebral artery
   _____ B. External jugular vein
   _____ C. Internal jugular vein
   _____ D. Common carotid artery

2. Which of the following is illustrated by 8?
   _____ A. Cricoid cartilage
   _____ B. Thyroid cartilage
   _____ C. Hyoid bone
   _____ D. Arytenoid cartilage

3. What number illustrates the laryngeal vestibule?
   _____ A. 8
   _____ B. 7
   _____ C. 1
   _____ D. 2

4. What number illustrates the piriform sinus?
   _____ A. 2
   _____ B. 7
   _____ C. 8
   _____ D. 1

5. Which of the following is illustrated by 7?
   _____ A. Vallecula
   _____ B. Thyroid cartilage
   _____ C. Laryngeal vestibule
   _____ D. Arytenoid cartilage

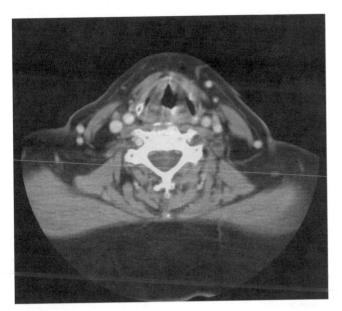
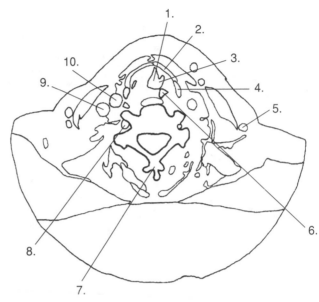

Figure 5-16

1. Which of the following is illustrated by 2?

    _____ A. Hyoid bone

    _____ B. Arytenoid cartilage

    _____ C. Cricoid cartilage

    _____ D. Thyroid cartilage

2. Which of the following is illustrated by 4?

    _____ A. Piriform sinus

    _____ B. Glottic space

    _____ C. Vallecula

    _____ D. Oropharynx

3. What number illustrates the arytenoid cartilage?

    _____ A. 6

    _____ B. 2

    _____ C. 3

    _____ D. 4

4. Which of the following is illustrated by 3?

    _____ A. Oropharynx

    _____ B. Glottic space

    _____ C. Arytenoid cartilage

    _____ D. Laryngeal vestibule

5. What number illustrates the common carotid artery?

    _____ A. 5

    _____ B. 8

    _____ C. 9

    _____ D. 10

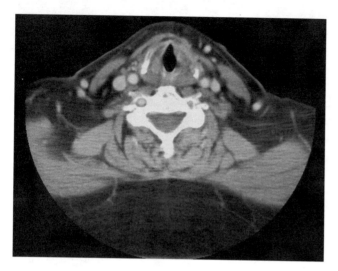

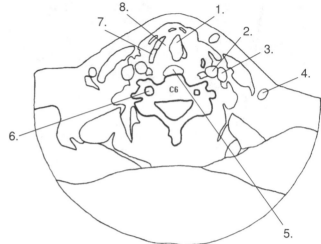

Figure 5-17

1. What number illustrates the vocal cord?

_____ A. 5

_____ B. 7

_____ C. 8

_____ D. 1

2. Which of the following is illustrated by 5?

_____ A. Esophagus

_____ B. Thyroid cartilage

_____ C. Infraglottic space

_____ D. Vocal cord

3. What number illustrates the common carotid artery?

_____ A. 6

_____ B. 2

_____ C. 3

_____ D. 4

4. Which of the following is illustrated by 1?

_____ A. Vocal cord

_____ B. Glottic space

_____ C. Vallecula

_____ D. Infraglottic space

5. Which of the following is illustrated by 7?

_____ A. Hyoid bone

_____ B. Thyroid cartilage

_____ C. Cricoid cartilage

_____ D. Vocal cord

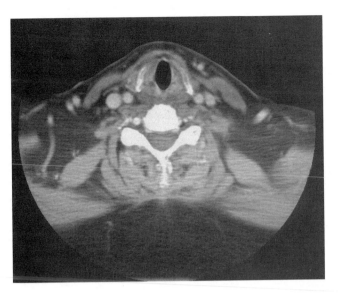 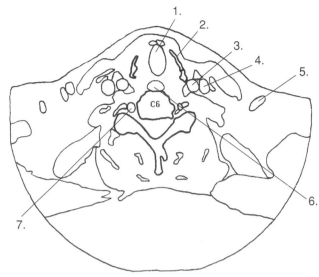

Figure 5-18

1. What number illustrates the common carotid artery?

   _____ A. 3

   _____ B. 7

   _____ C. 4

   _____ D. 5

2. Which of the following is illustrated by 2?

   _____ A. Trachea

   _____ B. Arytenoid cartilage

   _____ C. Cricoid cartilage

   _____ D. Thyroid cartilage

3. Which of the following is illustrated by 6?

   _____ A. Thyroid cartilage

   _____ B. Trachea

   _____ C. Esophagus

   _____ D. Laryngeal vestibule

4. What number illustrates the internal jugular vein?

   _____ A. 5

   _____ B. 7

   _____ C. 4

   _____ D. 3

5. Which of the following is illustrated by 1?

   _____ A. Trachea

   _____ B. Esophagus

   _____ C. Laryngeal vestibule

   _____ D. Laryngeal pharynx

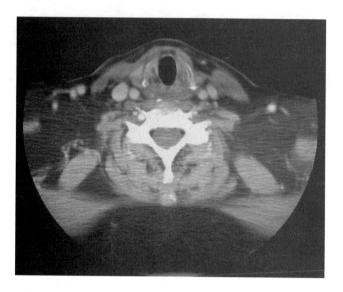 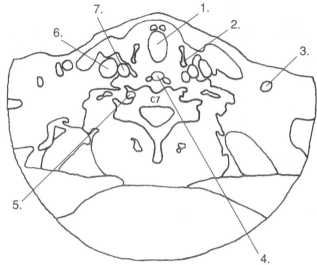

Figure 5-19

1. What number illustrates the internal jugular vein?

    _____ A. 7

    _____ B. 5

    _____ C. 6

    _____ D. 3

2. Which of the following is illustrated by 6?

    _____ A. External jugular vein

    _____ B. Internal jugular vein

    _____ C. Retromandibular vein

    _____ D. Subclavian vein

3. Which of the following is illustrated by 2?

    _____ A. Thyroid cartilage

    _____ B. Cricoid cartilage

    _____ C. Arytenoid cartilage

    _____ D. Hyoid bone

4. Which of the following is illustrated by 4?

    _____ A. Trachea

    _____ B. Esophagus

    _____ C. Vallecula

    _____ D. Laryngeal pharynx

5. Which of the following is illustrated by 1?

    _____ A. Vallecula

    _____ B. Esophagus

    _____ C. Laryngeal pharynx

    _____ D. Trachea

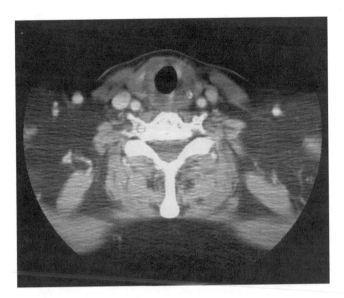
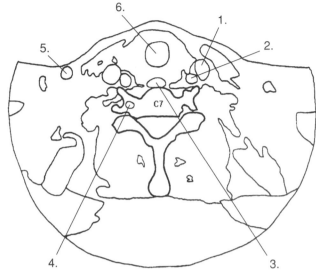

Figure 5-20

1. What number illustrates the common carotid artery?

    _____ A. 1

    _____ B. 3

    _____ C. 4

    _____ D. 2

2. Which of the following is illustrated by 1?

    _____ A. External jugular vein

    _____ B. Internal jugular vein

    _____ C. Vertebral vein

    _____ D. Common carotid artery

3. What number illustrates the external jugular vein?

    _____ A. 1

    _____ B. 4

    _____ C. 2

    _____ D. 5

4. Which of the following is illustrated by 6?

    _____ A. Infraglottic space

    _____ B. Glottic space

    _____ C. Trachea

    _____ D. Laryngeal vestibule

5. Which of the following is illustrated by 3?

    _____ A. Trachea

    _____ B. Laryngeal vestibule

    _____ C. Laryngeal pharynx

    _____ D. Esophagus

ANSWERS

Fig. 5-1

1. C
2. B
3. A
4. A
5. D

Fig. 5-2

1. A
2. B
3. D
4. A
5. C

Fig. 5-3

1. C
2. A
3. D
4. C
5. D

Fig. 5-4

1. B
2. A
3. C
4. D
5. C

Fig. 5-5

1. B
2. B
3. D
4. A
5. A

Fig. 5-6

1. C
2. D
3. A
4. D
5. C

Fig. 5-7

1. C
2. A
3. B
4. D
5. A

Fig. 5-8

1. C
2. B
3. A
4. C
5. D

Fig. 5-9

1. C
2. A
3. D
4. B
5. C

Fig. 5-10

1. A
2. C
3. D
4. C
5. A

Fig. 5-11

1. B
2. C
3. A
4. B
5. A

Fig. 5-12

1. D
2. C
3. B
4. A
5. C

Fig. 5-13

1. A
2. B
3. C
4. C
5. D

Fig. 5-14

1. C
2. B
3. C
4. A
5. D

Fig. 5-15

1. A
2. B
3. C
4. A
5. D

Fig. 5-16

1. D
2. A
3. A
4. B
5. D

Fig. 5-17

1. C
2. A
3. B
4. D
5. B

Fig. 5-18

1. A
2. D
3. C
4. C
5. A

Fig. 5-19

1. C
2. B
3. A
4. B
5. D

Fig. 5-20

1. D
2. B
3. D
4. C
5. D

Chapter 6

Spine

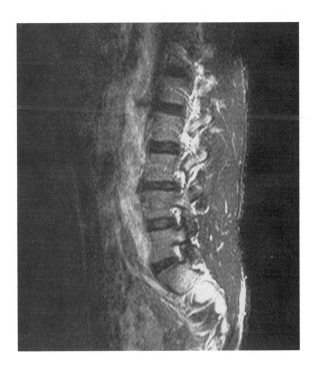
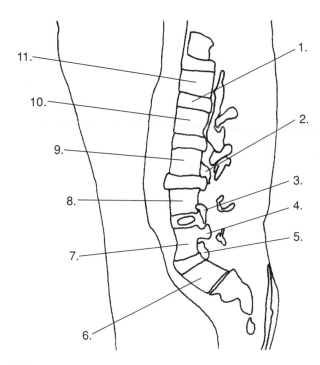

Figure 6-1

1. What number illustrates the pedicle?
   _____ A. 4
   _____ B. 2
   _____ C. 3
   _____ D. 5

2. Which of the following is illustrated by 8?
   _____ A. Vertebral body of L5
   _____ B. Intervertebral disk
   _____ C. Vertebral body of L4
   _____ D. L5 nerve roots

3. What number illustrates the vertebral body of S1?
   _____ A. 7
   _____ B. 8
   _____ C. 9
   _____ D. 6

4. Which of the following is illustrated by 1?
   _____ A. L3 nerve roots
   _____ B. Intervertebral disk
   _____ C. Vertebral body of L1
   _____ D. Pedicle of L5

5. What number illustrates the L5 nerve roots?
   _____ A. 6
   _____ B. 4
   _____ C. 5
   _____ D. 3

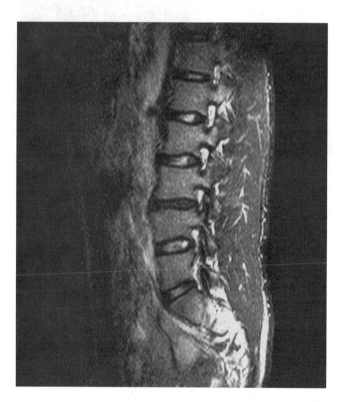
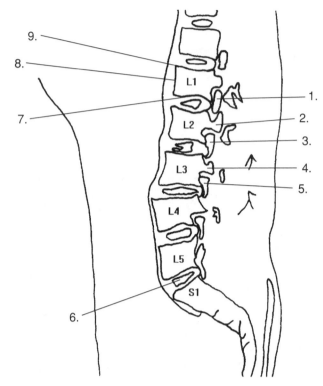

Figure 6-2

1. What number illustrates the superior vertebral end plate?

      _____ A. 1

      _____ B. 7

      _____ C. 2

      _____ D. 9

2. Which of the following is illustrated by 5?

      _____ A. L3 nerve roots

      _____ B. L2 nerve roots

      _____ C. L4 nerve roots

      _____ D. L3 pedicle

3. What number illustrates the L2 pedicle?

      _____ A. 4

      _____ B. 2

      _____ C. 5

      _____ D. 7

4. Which of the following is illustrated by 8?

      _____ A. Inferior vertebral endplate L1

      _____ B. L1 nerve roots

      _____ C. Anterior cortical bone of body

      _____ D. L2 nerve roots

5. Which of the following is illustrated by 6?

      _____ A. Anterior cortical bone of body

      _____ B. L3 pedicle

      _____ C. Intervertebral disk

      _____ D. Inferior vertebral endplate L1

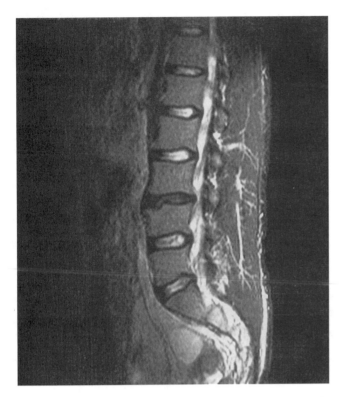
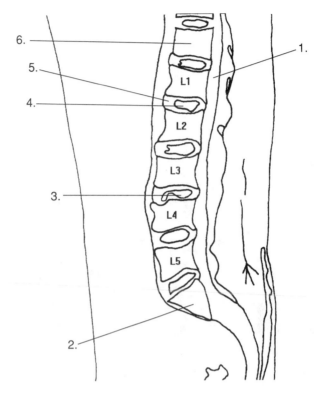

Figure 6-3

1. Which of the following is illustrated by 1?

    _____ A. Spinal cord

    _____ B. Edge of dural sac

    _____ C. Anulus fibrosus

    _____ D. Cauda equina

2. What number illustrates the herniated disk?

    _____ A. 4

    _____ B. 2

    _____ C. 5

    _____ D. 3

3. What number illustrates the nucleus pulposus?

    _____ A. 4

    _____ B. 6

    _____ C. 1

    _____ D. 5

4. Which of the following is illustrated by 6?

    _____ A. L1 vertebral body

    _____ B. L2 vertebral body

    _____ C. T12 vertebral body

    _____ D. Intervertebral disk

5. What number illustrates the anulus fibrosus?

    _____ A. 5

    _____ B. 1

    _____ C. 3

    _____ D. 4

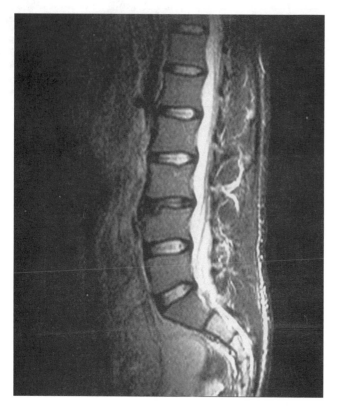 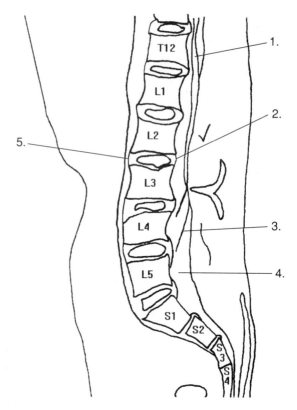

Figure 6-4

1. Which of the following is illustrated by 1?

_____ A. Spinal cord

_____ B. Edge of dural sac

_____ C. Anulus fibrosus

_____ D. Cauda equina

2. Which of the following is illustrated by 2?

_____ A. Anterior longitudinal ligament

_____ B. Posterior longitudinal ligament

_____ C. Herniated disk

_____ D. Nucleus pulposus

3. What number illustrates the cauda equina?

_____ A. 1

_____ B. 3

_____ C. 4

_____ D. 2

4. What number illustrates the subarachnoid space?

_____ A. 1

_____ B. 2

_____ C. 3

_____ D. 4

5. Which of the following is illustrated by 5?

_____ A. Anterior longitudinal ligament

_____ B. Posterior longitudinal ligament

_____ C. Herniated disk

_____ D. Nucleus pulposus

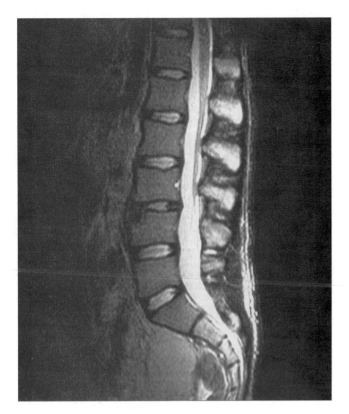
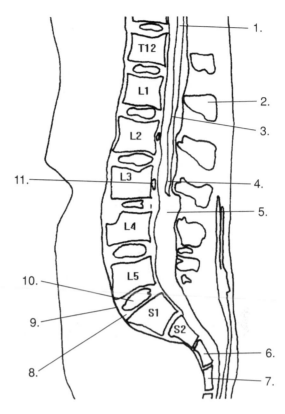

Figure 6-5

1. What number illustrates the conus medullaris?
    _____ A. 1
    _____ B. 5
    _____ C. 4
    _____ D. 3

2. Which of the following is illustrated by 10?
    _____ A. Intervertebral disk L4-L5
    _____ B. Nucleus pulposus
    _____ C. Anulus fibrosus
    _____ D. Intervertebral disk S1-S2

3. What number illustrates the spinal cord?
    _____ A. 5
    _____ B. 3
    _____ C. 1
    _____ D. 4

4. Which of the following is illustrated by 5?
    _____ A. Subarachnoid space
    _____ B. Epidural space
    _____ C. L1 spinous process
    _____ D. Subdural space

5. Which of the following is illustrated by 4?
    _____ A. Spinal cord
    _____ B. Conus medullaris
    _____ C. Cauda equina
    _____ D. Basivertebral vein

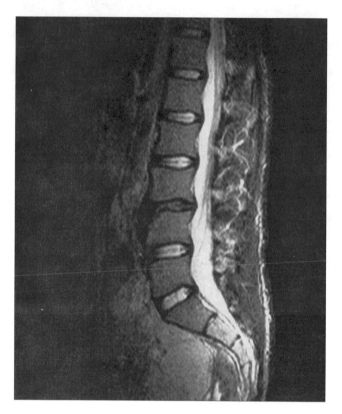
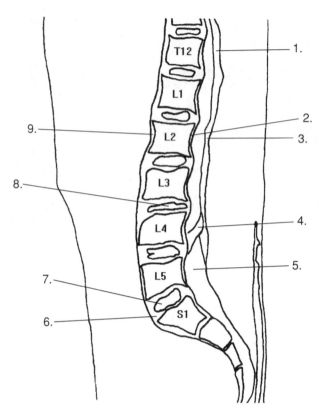

Figure 6-6

1. What number illustrates the cauda equina?

   _____ A. 1

   _____ B. 4

   _____ C. 3

   _____ D. 5

2. Which of the following is illustrated by 7?

   _____ A. Anulus fibrosus

   _____ B. Ruptured disk

   _____ C. Nucleus pulposus

   _____ D. Intervertebral disk L4-L5

3. Which of the following is illustrated by 2?

   _____ A. Anterior longitudinal ligament

   _____ B. Conus medullaris

   _____ C. Spinal cord

   _____ D. Posterior longitudinal ligament

4. Which of the following is illustrated by 9?

   _____ A. Anterior longitudinal ligament

   _____ B. Anulus fibrosus

   _____ C. Subarachnoid space

   _____ D. Posterior longitudinal ligament

5. What number illustrates the ruptured disk?

   _____ A. 7

   _____ B. 9

   _____ C. 5

   _____ D. 8

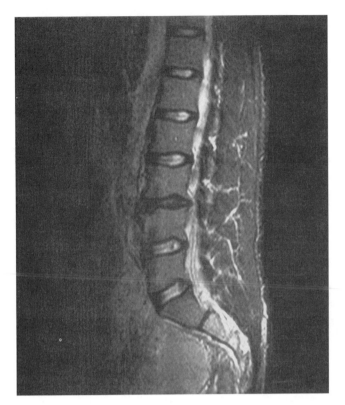
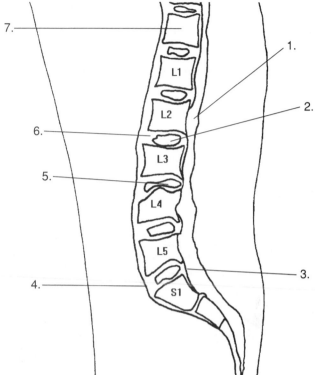

Figure 6-7

1. Which of the following is illustrated by 2?

    _____ A. Anulus fibrosus

    _____ B. Herniated disk

    _____ C. Nucleus pulposus

    _____ D. Edge of dural sac

2. What number illustrates the anterior longitudinal ligament?

    _____ A. 4

    _____ B. 3

    _____ C. 5

    _____ D. 2

3. Which of the following is illustrated by 6?

    _____ A. Herniated disk

    _____ B. Anulus fibrosus

    _____ C. Nucleus pulposus

    _____ D. Edge of dural sac

4. Which of the following is illustrated by 1?

    _____ A. Edge of the dural sac

    _____ B. Epidural space

    _____ C. Basivertebral vein

    _____ D. Pia mater

5. What number illustrates the herniated disk?

    _____ A. 2

    _____ B. 6

    _____ C. 4

    _____ D. 5

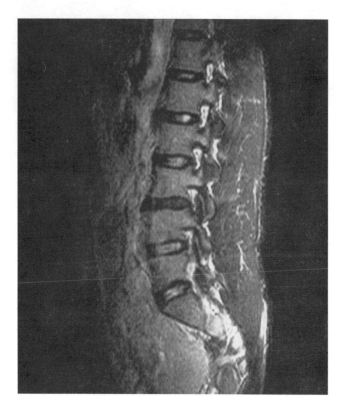
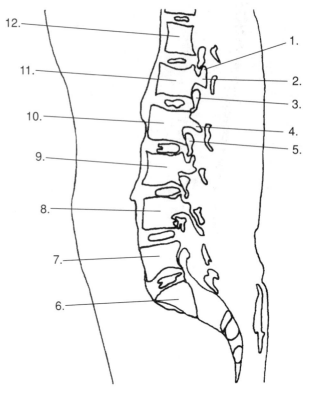

Figure 6-8

1. What number illustrates the L1 nerve roots?
    _____ A. 2
    _____ B. 5
    _____ C. 3
    _____ D. 1

2. Which of the following is illustrated by 10?
    _____ A. L2
    _____ B. L1
    _____ C. L3
    _____ D. T1

3. What number illustrates the L1 pedicle?
    _____ A. 4
    _____ B. 3
    _____ C. 5
    _____ D. 2

4. Which of the following is illustrated by 8?
    _____ A. L3
    _____ B. L4
    _____ C. S1
    _____ D. L5

5. Which of the following is illustrated by 3?
    _____ A. L2 nerve roots
    _____ B. L1 pedicle
    _____ C. L1 nerve roots
    _____ D. L2 pedicle

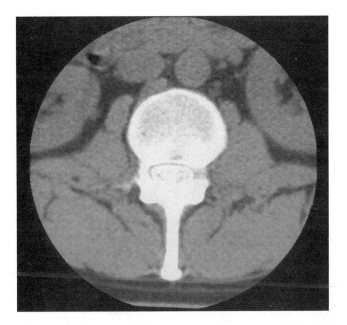
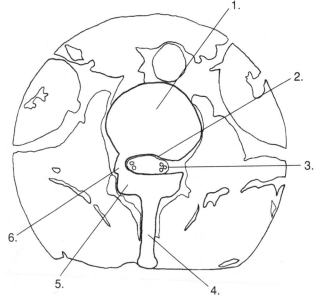

Figure 6-9

1. What number illustrates the epidural space?

    _____ A. 3

    _____ B. 2

    _____ C. 5

    _____ D. 6

2. Which of the following is illustrated by 4?

    _____ A. Lamina of L2

    _____ B. Pedicle of L2

    _____ C. L2 vertebral body

    _____ D. Spinous process of L2

3. What number illustrates the lamina of L2?

    _____ A. 5

    _____ B. 2

    _____ C. 6

    _____ D. 4

4. Which of the following is illustrated by 5?

    _____ A. Pedicle of L2

    _____ B. Lamina of L2

    _____ C. Spinous process of L2

    _____ D. Epidural space

5. Which of the following is illustrated by 3?

    _____ A. Spinal cord

    _____ B. Epidural space

    _____ C. Dural sac

    _____ D. Subarachnoid space

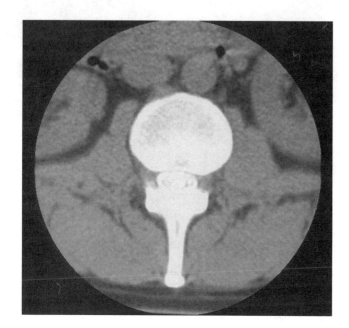 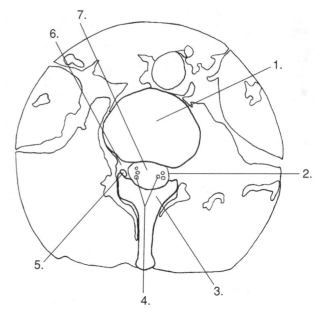

Figure 6-10

1. Which of the following is illustrated by 4?

_____ A. Posterior dorsal root ganglion

_____ B. Epidural space

_____ C. Cauda equina

_____ D. Conus medullaris

2. Which of the following is illustrated by 2?

_____ A. Dural sac

_____ B. Posterior dorsal root ganglion

_____ C. Subarachnoid space

_____ D. Cauda equina

3. Which of the following is illustrated by 6?

_____ A. Vertebral foramen

_____ B. Posterior (dorsal) root ganglion

_____ C. Cauda equina

_____ D. Intervertebral foramen

4. Which of the following is illustrated by 7?

_____ A. Spinal cord

_____ B. Subarachnoid space

_____ C. Posterior dorsal root ganglion

_____ D. Epidural space

5. Which of the following is illustrated by 5?

_____ A. Cauda equina

_____ B. Conus medullaris

_____ C. Posterior (dorsal) root ganglion

_____ D. Anterior nerve root

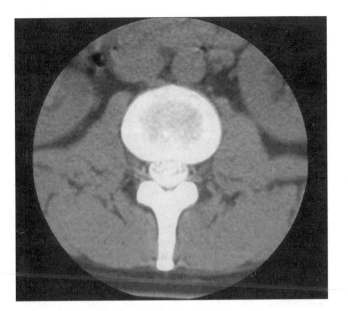
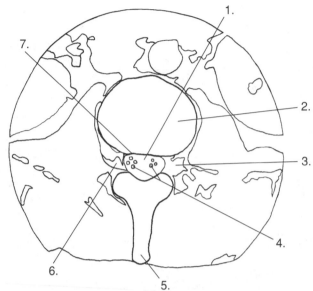

Figure 6-11

1. Which of the following is illustrated by 2?

_____ A. Spinous process

_____ B. Superior endplate

_____ C. Inferior endplate

_____ D. Intervertebral disk

2. What number illustrates the cauda equina?

_____ A. 4

_____ B. 7

_____ C. 6

_____ D. 1

3. Which of the following is illustrated by 7?

_____ A. Pia mater

_____ B. Arachnoid mater

_____ C. Subarachnoid space

_____ D. Dural sac

4. Which of the following is illustrated by 3?

_____ A. Intervertebral disk

_____ B. Intervertebral foramen

_____ C. Dural sac

_____ D. Cauda equina

5. What number illustrates the subarachnoid space?

_____ A. 7

_____ B. 4

_____ C. 1

_____ D. 6

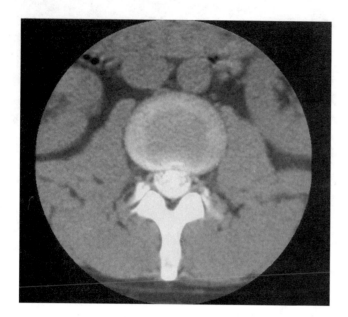
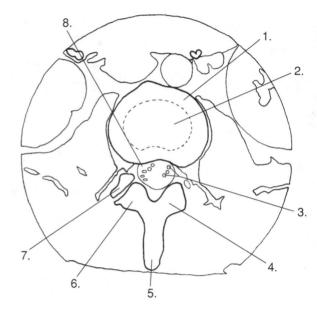

Figure 6-12

1. Which of the following is illustrated by 6?

_____ A. Spinous process of L2

_____ B. Lamina of L2

_____ C. Superior articular process of L3

_____ D. Inferior articular process of L2

2. What number illustrates the anulus fibrosus?

_____ A. 8

_____ B. 1

_____ C. 4

_____ D. 2

3. Which of the following is illustrated by 2?

_____ A. Nucleus pulposus

_____ B. Anulus fibrosus

_____ C. Vertebral body of L3

_____ D. Vertebral body of L2

4. What number illustrates the superior articular process of L3?

_____ A. 5

_____ B. 4

_____ C. 7

_____ D. 6

5. Which of the following is illustrated by 8?

_____ A. Spinal cord

_____ B. Subarachnoid space

_____ C. Cauda equina

_____ D. Conus medullaris

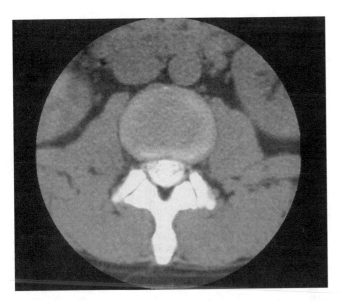
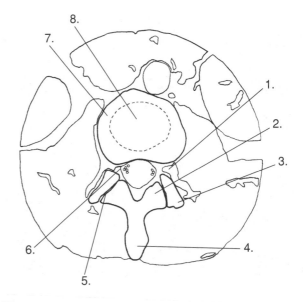

Figure 6-13

1. Which of the following is illustrated by 5?
    _____ A. Zygapophysis
    _____ B. Inferior articular process of L2
    _____ C. Lamina of L2
    _____ D. Epidural space

2. Which of the following is illustrated by 3?
    _____ A. Inferior articular process of L2
    _____ B. Lamina of L2
    _____ C. Superior articular process of L3
    _____ D. Pedicle of L2

3. What number illustrates the anulus fibrosus?
    _____ A. 8
    _____ B. 7
    _____ C. 6
    _____ D. 2

4. Which of the following is illustrated by 6?
    _____ A. Dural sac
    _____ B. Zygapophysis
    _____ C. Nucleus pulposus
    _____ D. Cauda equina

5. What number illustrates the inferior articular process of L2?
    _____ A. 1
    _____ B. 3
    _____ C. 6
    _____ D. 2

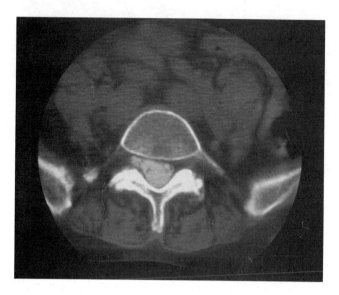 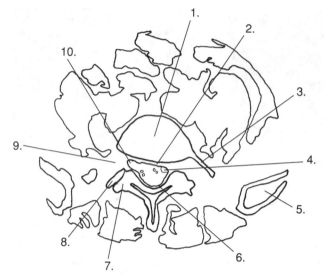

Figure 6-14

1. Which of the following is illustrated by 6?

    _____ A. Spinal cord

    _____ B. Dural sac

    _____ C. Conus medullaris

    _____ D. S1 nerve root

2. Which of the following is illustrated by 10?

    _____ A. Intervertebral foramen

    _____ B. Conus medullaris

    _____ C. Cauda equina

    _____ D. L5 nerve root

3. What number illustrates the epidural space?

    _____ A. 2

    _____ B. 7

    _____ C. 4

    _____ D. 9

4. Which of the following is illustrated by 9?

    _____ A. L5 nerve root

    _____ B. Posterior (dorsal) root ganglion

    _____ C. Intervertebral foramen

    _____ D. Dural sac

5. Which of the following is illustrated by 3?

    _____ A. Transverse process of L5

    _____ B. Lateral part of S1

    _____ C. Lamina of S1

    _____ D. Pedicle of S1

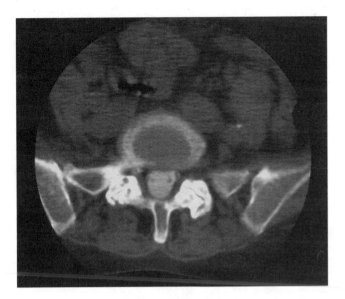
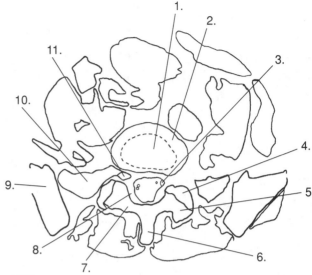

Figure 6-15

1. What number illustrates the lamina of L5?

    _____ A. 8

    _____ B. 7

    _____ C. 10

    _____ D. 5

2. Which of the following is illustrated by 3?

    _____ A. Spinal cord

    _____ B. Cauda equina

    _____ C. Epidural space

    _____ D. S1 nerve root

3. What number illustrates the nucleus pulposus?

    _____ A. 11

    _____ B. 8

    _____ C. 1

    _____ D. 2

4. What number illustrates the lateral part of S1?

    _____ A. 10

    _____ B. 4

    _____ C. 5

    _____ D. 9

5. Which of the following is illustrated by 5?

    _____ A. Superior articular process of S1

    _____ B. Superior articular process of L5

    _____ C. Inferior articular process of S1

    _____ D. Inferior articular process of L5

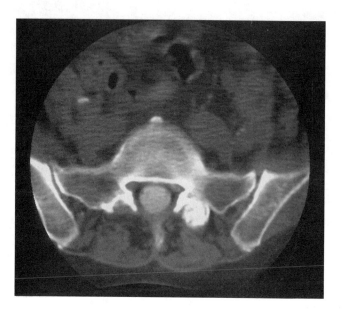
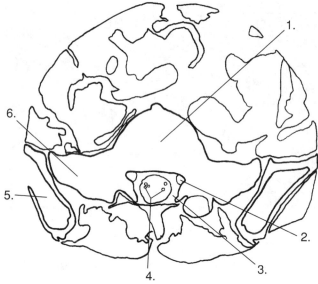

Figure 6-16

1. Which of the following is illustrated by 3?

    _____ A. S1 nerve root

    _____ B. Epidural space

    _____ C. Subarachnoid space

    _____ D. S2-S5 nerve roots

2. What number illustrates the ilium?

    _____ A. 6

    _____ B. 1

    _____ C. 2

    _____ D. 5

3. Which of the following is illustrated by 1?

    _____ A. Inferior endplate of L5

    _____ B. Intervertebral disk

    _____ C. Superior endplate of S1

    _____ D. Zygapophysis

4. Which of the following is illustrated by 6?

    _____ A. Lateral part of S1

    _____ B. Ilium

    _____ C. Transverse process of L5

    _____ D. Superior endplate of S1

5. Which of the following is illustrated by 4?

    _____ A. S1 nerve root

    _____ B. S2-S5 nerve roots

    _____ C. Dural sac

    _____ D. Epidural space

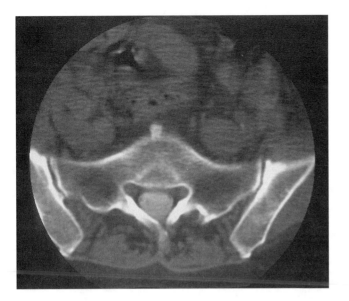 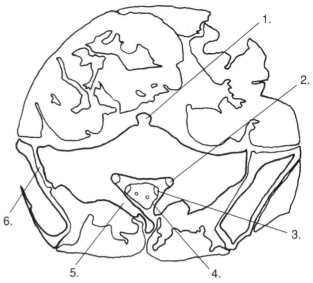

Figure 6-17

1. What number illustrates the dural sac?

_____ A. 3

_____ B. 6

_____ C. 2

_____ D. 4

2. Which of the following is illustrated by 2?

_____ A. S1 nerve root

_____ B. Subarachnoid space

_____ C. Vertebral artery

_____ D. Epidural space

3. Which of the following is illustrated by 6?

_____ A. Lamina of S1

_____ B. Zygapophysis

_____ C. Sacroiliac joint

_____ D. L5- S1 intervertebral joint

4. What number illustrates the osteophyte?

_____ A. 5

_____ B. 1

_____ C. 6

_____ D. 2

5. Which of the following is illustrated by 3?

_____ A. Dural sac

_____ B. Epidural space

_____ C. Spinal cord

_____ D. S2 nerve root

ANSWERS

Fig. 6-1

1. A
2. C
3. D
4. B
5. C

Fig. 6-2

1. D
2. A
3. B
4. C
5. C

Fig. 6-3

1. B
2. D
3. A
4. C
5. A

Fig. 6-4

1. A
2. B
3. B
4. D
5. A

Fig. 6-5

1. D
2. B
3. C
4. A
5. C

Fig. 6-6

1. B
2. C
3. D
4. A
5. D

Fig. 6-7

1. C
2. A
3. B
4. A
5. D

Fig. 6-8

1. C
2. A
3. D
4. B
5. C

Fig. 6-9

1. B
2. D
3. A
4. B
5. C

Fig. 6-10

1. C
2. A
3. D
4. B
5. C

Fig. 6-11

1. C
2. A
3. D
4. B
5. C

Fig. 6-12

1. D
2. B
3. A
4. C
5. B

Fig. 6-13

1. A
2. C
3. B
4. A
5. D

Fig. 6-14

1. B
2. D
3. A
4. C
5. A

Fig. 6-15

1. B
2. D
3. C
4. A
5. D

Fig. 6-16

1. B
2. D
3. C
4. A
5. B

Fig. 6-17

1. D
2. A
3. C
4. B
5. D

Chapter 7

Joints

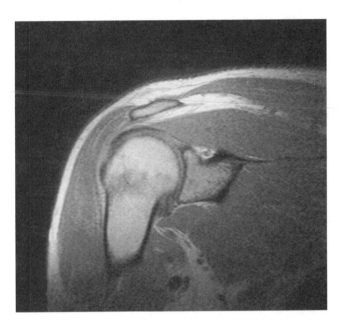

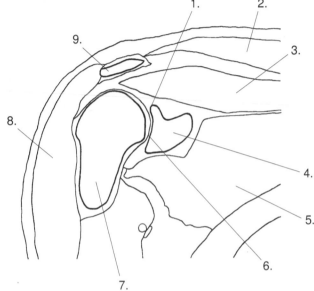

Figure 7-1

1. Which of the following illustrates 6?
   _____ A. Glenohumeral joint
   _____ B. Acromioclavicular joint
   _____ C. Axillary recess
   _____ D. Inferior glenoid labrum

2. What number illustrates the trapezius muscle?
   _____ A. 3
   _____ B. 2
   _____ C. 5
   _____ D. 8

3. Which of the following is illustrated by 4?
   _____ A. Superior glenoid labrum
   _____ B. Glenohumeral joint
   _____ C. Acromion process of scapula
   _____ D. Glenoid process of scapula

4. Which of the following is illustrated by 1?
   _____ A. Superior glenoid labrum
   _____ B. Acromion process
   _____ C. Axillary recess
   _____ D. Acromioclavicular joint

5. What number illustrates the supraspinatus muscle?
   _____ A. 2
   _____ B. 5
   _____ C. 3
   _____ D. 8

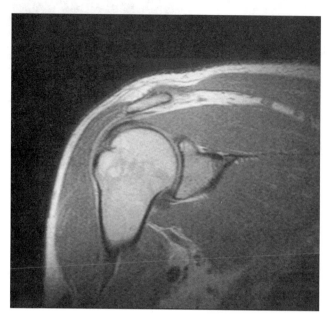
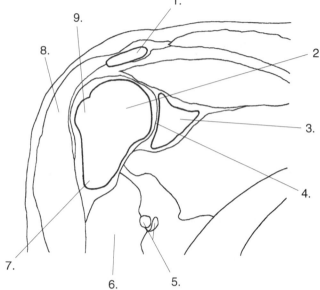

Figure 7-2

1. Which of the following is illustrated by 6?
    _____ A. Subscapular muscle
    _____ B. Teres minor muscle
    _____ C. Deltoid muscle
    _____ D. Teres major muscle

2. Which of the following is illustrated by 9?
    _____ A. Lesser tubercle of the humerus
    _____ B. Greater tubercle of the humerus
    _____ C. Surgical neck
    _____ D. Head of the humerus

3. Which of the following is illustrated by 1?
    _____ A. Greater tubercle of the humerus
    _____ B. Surgical neck of the humerus
    _____ C. Acromion process of the scapula
    _____ D. Head of the humerus

4. Which of the following is illustrated by 7?
    _____ A. Surgical neck of the humerus
    _____ B. Greater tubercle of the humerus
    _____ C. Head of the humerus
    _____ D. Acromion process of the scapula

5. Which of the following is illustrated by 2?
    _____ A. Acromion process of the scapula
    _____ B. Greater tubercle of the humerus
    _____ C. Head of the humerus
    _____ D. Surgical neck of the humerus

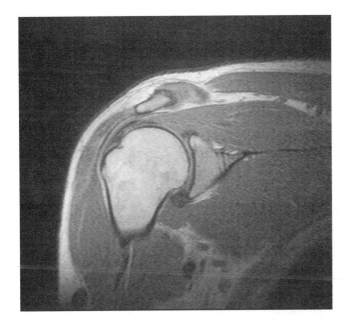
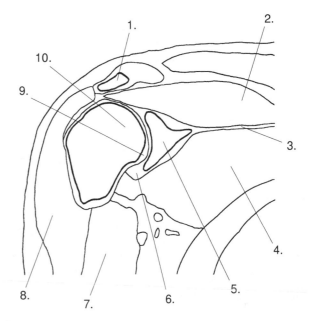

Figure 7-3

1. Which of the following is illustrated by 1?

    _____ A. Glenoid process of scapula

    _____ B. Clavicle

    _____ C. Greater tubercle

    _____ D. Acromion process of scapula

2. What number illustrates the supraspinatus muscle?

    _____ A. 2

    _____ B. 8

    _____ C. 7

    _____ D. 4

3. Which of the following is illustrated by 6?

    _____ A. Axillary recess of the articular cavity

    _____ B. Glenoid process of the scapula

    _____ C. Acromioclavicular joint

    _____ D. Articular cartilage

4. What number illustrates the articular cartilage?

    _____ A. 3

    _____ B. 10

    _____ C. 9

    _____ D. 6

5. Which of the following is illustrated by 3?

    _____ A. Articular cartilage

    _____ B. Glenoid labrum

    _____ C. Glenoid process of scapula

    _____ D. Body of scapula

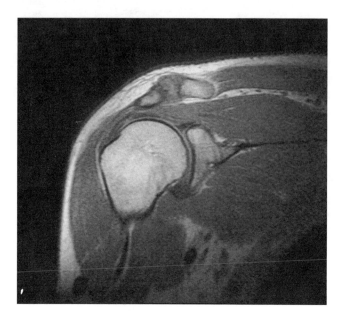 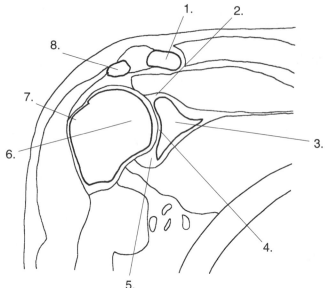

Figure 7-4

1. Which of the following is illustrated by 2?

_____ A. Articular cartilage in the glenoid fossa

_____ B. Superior glenoid labrum

_____ C. Inferior glenoid labrum

_____ D. Axillary recess of articular cavity

2. Which of the following is illustrated by 1?

_____ A. Acromion process of scapula

_____ B. Glenoid process of scapula

_____ C. Superior glenoid labrum

_____ D. Clavicle

3. What number illustrates the articular cartilage in the glenoid fossa?

_____ A. 4

_____ B. 3

_____ C. 2

_____ D. 5

4. What number illustrates the inferior glenoid labrum?

_____ A. 4

_____ B. 3

_____ C. 5

_____ D. 2

5. Which of the following is illustrated by 3?

_____ A. Superior glenoid labrum

_____ B. Inferior glenoid labrum

_____ C. Glenoid process of the scapula

_____ D. Acromion process of the scapula

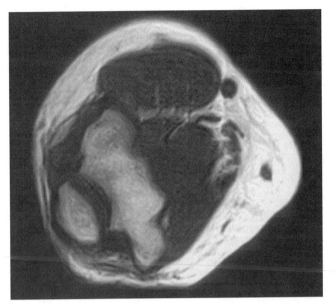
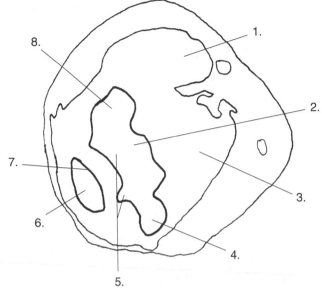

Figure 7-5

1. What number illustrates the trochlear notch?

_____ A. 4

_____ B. 7

_____ C. 6

_____ D. 5

2. Which of the following is illustrated by 2?

_____ A. Radius

_____ B. Ulna

_____ C. Trochlea

_____ D. Humerus

3. What number illustrates the olecranon process of the ulna?

_____ A. 7

_____ B. 2

_____ C. 6

_____ D. 5

4. Which of the following is illustrated by 4?

_____ A. Medial epicondyle

_____ B. Capitulum

_____ C. Trochlea

_____ D. Lateral epicondyle

5. Which of the following is illustrated by 1?

_____ A. Biceps muscle

_____ B. Triceps muscle

_____ C. Brachialis muscle

_____ D. Brachioradialis muscle

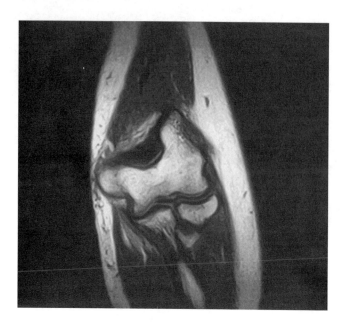 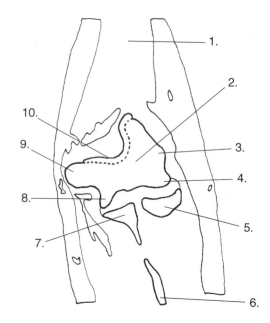

Figure 7-6

1. Which of the following is illustrated by 8?
   _____ A. Coronoid process of ulna
   _____ B. Medial epicondyle
   _____ C. Trochlea
   _____ D. Capitulum

2. What number illustrates the capitulum?
   _____ A. 4
   _____ B. 2
   _____ C. 5
   _____ D. 3

3. Which of the following is illustrated by the 10?
   _____ A. Capitulum
   _____ B. Lateral epicondyle
   _____ C. Trochlea
   _____ D. Olecranon fossa

4. What number illustrates the lateral epicondyle?
   _____ A. 7
   _____ B. 3
   _____ C. 5
   _____ D. 9

5. What number illustrates the coronoid process of the ulna?
   _____ A. 5
   _____ B. 6
   _____ C. 9
   _____ D. 7

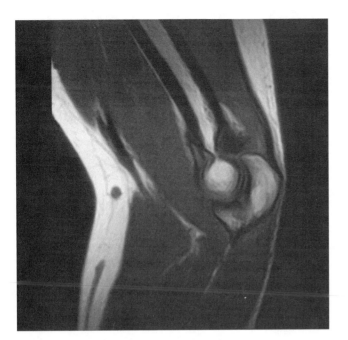
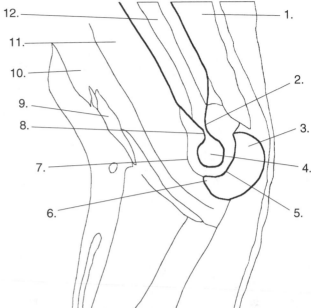

Figure 7-7

1. What number illustrates the olecranon fossa?

_____ A. 8

_____ B. 5

_____ C. 7

_____ D. 2

2. Which of the following is illustrated by 5?

_____ A. Olecranon process of ulna

_____ B. Articular capsule

_____ C. Trochlear notch

_____ D. Coronoid process of ulna

3. What number illustrates the triceps muscle?

_____ A. 1

_____ B. 12

_____ C. 11

_____ D. 10

4. Which of the following is illustrated by 9?

_____ A. Brachialis muscle

_____ B. Brachial artery

_____ C. Biceps muscle

_____ D. Axillary artery

5. Which of the following is illustrated by 7?

_____ A. Capitulum

_____ B. Coronoid fossa

_____ C. Articular capsule

_____ D. Trochlear notch

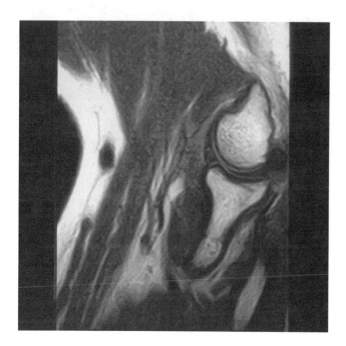 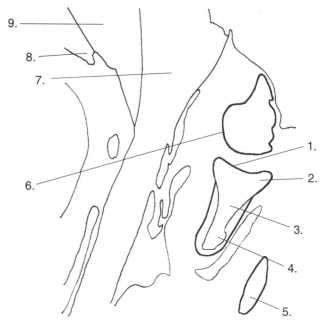

Figure 7-8

1. Which of the following is illustrated by 3?

    _____ A. Fovea of the radius

    _____ B. Neck of radius

    _____ C. Radial tuberosity

    _____ D. Head of radius

2. What number illustrates the biceps muscle?

    _____ A. 8

    _____ B. 6

    _____ C. 9

    _____ D. 7

3. Which of the following is illustrated by 7?

    _____ A. Biceps muscle

    _____ B. Deltoid muscle

    _____ C. Brachialis muscle

    _____ D. Brachioradialis muscle

4. What number illustrates the fovea of the radius?

    _____ A. 2

    _____ B. 5

    _____ C. 1

    _____ D. 6

5. What number illustrates the capitulum?

    _____ A. 6

    _____ B. 2

    _____ C. 1

    _____ D. 5

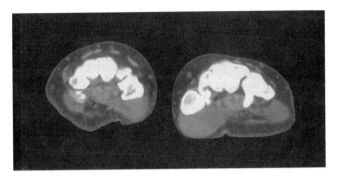 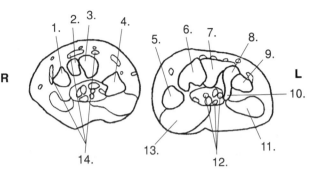

Figure 7-9

1. Which of the following is illustrated by 9?
   _____ A. Trapezoid
   _____ B. Triquetrum
   _____ C. Trapezium
   _____ D. Hamate

2. What number illustrates the trapezoid?
   _____ A. 4
   _____ B. 2
   _____ C. 7
   _____ D. 3

3. Which of the following is illustrated by 6?
   _____ A. Trapezium
   _____ B. Hamate
   _____ C. Trapezoid
   _____ D. Capitate

4. What number illustrates the abductor digiti minimi muscle?
   _____ A. 11
   _____ B. 13
   _____ C. 12
   _____ D. 14

5. What number illustrates the flexor digitorum superficialis and profundus tendons?
   _____ A. 14
   _____ B. 13
   _____ C. 11
   _____ D. 12

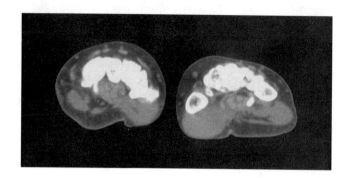 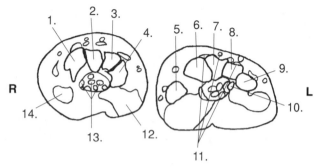

Figure 7-10

1. What number illustrates the capitate?

   _____ A. 8

   _____ B. 6

   _____ C. 3

   _____ D. 7

2. Which of the following is illustrated by 10?

   _____ A. Triquetrum

   _____ B. Hamulus of hamate

   _____ C. Pisiform

   _____ D. Scaphoid

3. What number illustrates the right hamate?

   _____ A. 3

   _____ B. 1

   _____ C. 4

   _____ D. 2

4. Which of the following is illustrated by 11?

   _____ A. Extensor tendons

   _____ B. Flexor digitorum superficialis and profundus tendons

   _____ C. Abductor digiti minimi muscle

   _____ D. Abductor pollicis brevis and opponens pollicis muscles

5. Which of the following is illustrated by 13?

   _____ A. Extensor tendons

   _____ B. Flexor digitorum superficialis and profundus tendons

   _____ C. Abductor digiti minimi muscle

   _____ D. Abductor pollicis brevis and opponens pollicis muscles

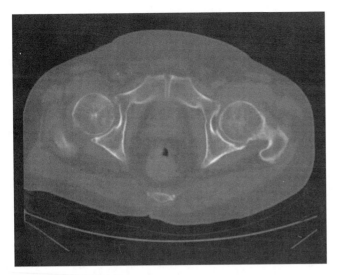 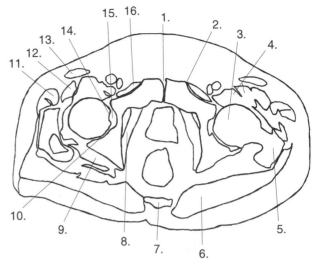

Figure 7-11

1. What number illustrates the sartorius muscle?

    _____ A. 14

    _____ B. 13

    _____ C. 12

    _____ D. 11

2. Which of the following is illustrated by 14?

    _____ A. Acetabular fossa

    _____ B. Lunate surface of acetabulum

    _____ C. Fovea capitis femoris

    _____ D. Pubic symphysis

3. What number illustrates the tensor fascia latae muscle?

    _____ A. 11

    _____ B. 14

    _____ C. 12

    _____ D. 13

4. Which of the following is illustrated by 1?

    _____ A. Pubic symphysis

    _____ B. Acetabular fossa

    _____ C. Obturator foramen

    _____ D. Sacroiliac joint

5. What number illustrates the superior gamellus muscle?

    _____ A. 9

    _____ B. 8

    _____ C. 6

    _____ D. 4

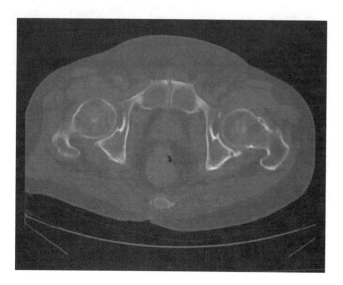 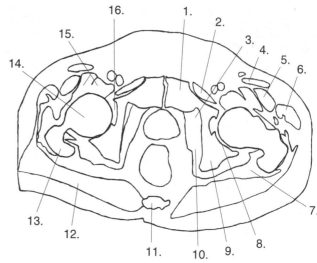

Figure 7-12

1. Which of the following is illustrated by 3?

    _____ A. Lunate surface of acetabulum

    _____ B. Obturator foramen

    _____ C. Pubic symphysis

    _____ D. Acetabular fossa

2. Which of the following is illustrated by 16?

    _____ A. Obturator foramen

    _____ B. Lunate surface of acetabulum

    _____ C. Fovea capitis femoris

    _____ D. Acetabular fossa

3. What number illustrates the iliopsoas muscle?

    _____ A. 5

    _____ B. 10

    _____ C. 7

    _____ D. 15

4. What number illustrates the obturator internus muscle?

    _____ A. 10

    _____ B. 15

    _____ C. 7

    _____ D. 2

5. What number illustrates the pectineus muscle?

    _____ A. 10

    _____ B. 15

    _____ C. 7

    _____ D. 2

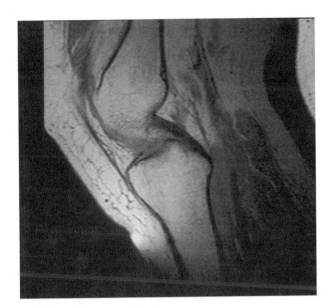 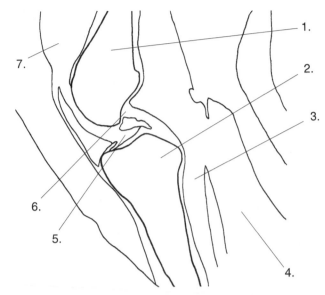

Figure 7-13

1. Which of the following is illustrated by 6?

    _____ A. Posterior cruciate ligament

    _____ B. Popliteus muscle

    _____ C. Anterior cruciate ligament

    _____ D. Tibia

2. Which of the following is illustrated by 2?

    _____ A. Femur

    _____ B. Fibula

    _____ C. Patella

    _____ D. Tibia

3. What number illustrates the popliteus muscle?

    _____ A. 4

    _____ B. 7

    _____ C. 5

    _____ D. 3

4. Which of the following is illustrated by 5?

    _____ A. Medial collateral ligament

    _____ B. Anterior cruciate ligament

    _____ C. Posterior cruciate ligament

    _____ D. Popliteus muscle

5. What number illustrates the quadriceps muscle?

    _____ A. 7

    _____ B. 4

    _____ C. 6

    _____ D. 3

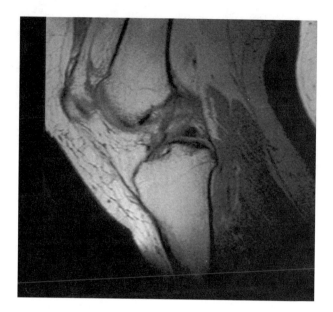

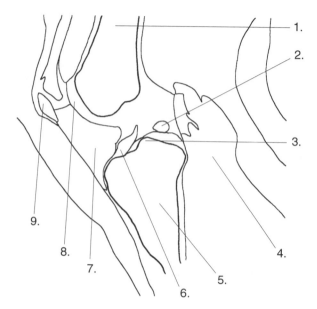

Figure 7-14

1. Which of the following is illustrated by 9?

    _____ A. Articular cartilage

    _____ B. Infrapatellar fat pad

    _____ C. Anterior cruciate ligament

    _____ D. Medial patella

2. Which of the following is illustrated by 3?

    _____ A. Posterior cruciate ligament

    _____ B. Intercondylar eminence

    _____ C. Anterior cruciate ligament

    _____ D. Lateral head of gastrocnemius muscle

3. What number illustrates the anterior cruciate ligament?

    _____ A. 6

    _____ B. 8

    _____ C. 3

    _____ D. 2

4. Which of the following is illustrated by 7?

    _____ A. Patellar tendon

    _____ B. Patella

    _____ C. Infrapatellar fat pad

    _____ D. Quadriceps tendon

5. What number illustrates the articular cartilage?

    _____ A. 8

    _____ B. 3

    _____ C. 2

    _____ D. 6

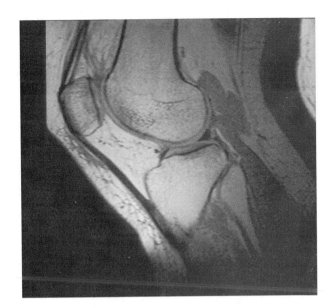 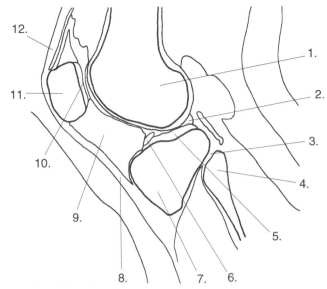

Figure 7-15

1. What number illustrates the posterior horn of the lateral meniscus?

   _____ A. 5

   _____ B. 2

   _____ C. 7

   _____ D. 6

2. What number illustrates the femoropatellar joint?

   _____ A. 5

   _____ B. 3

   _____ C. 9

   _____ D. 10

3. Which of the following is illustrated by 5?

   _____ A. Head of fibula

   _____ B. Anterior horn of lateral meniscus

   _____ C. Posterior horn of lateral meniscus

   _____ D. Articular cartilage

4. What number illustrates the patella?

   _____ A. 11

   _____ B. 8

   _____ C. 9

   _____ D. 1

5. Which of the following is illustrated by 6?

   _____ A. Anterior horn of medial meniscus

   _____ B. Anterior cruciate ligament

   _____ C. Posterior cruciate ligament

   _____ D. Anterior horn of lateral meniscus

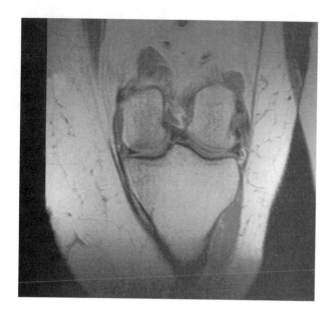
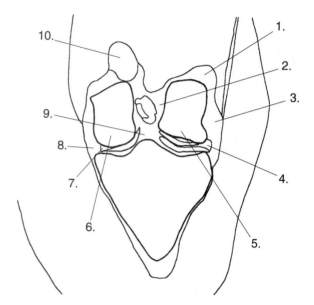

Figure 7-16

1. Which of the following is illustrated by 4?

    _____ A. Posterior cruciate ligament

    _____ B. Lateral collateral ligament

    _____ C. Lateral meniscus

    _____ D. Articular cartilage

2. What number illustrates the posterior cruciate ligament?

    _____ A. 6

    _____ B. 2

    _____ C. 5

    _____ D. 9

3. What number illustrates the medial meniscus?

    _____ A. 7

    _____ B. 9

    _____ C. 8

    _____ D. 4

4. Which of the following is illustrated by 1?

    _____ A. Lateral head of gastrocnemius muscle

    _____ B. Popliteal tendon

    _____ C. Quadriceps muscle

    _____ D. Medial head of gastrocnemius muscle

5. Which of the following is illustrated by 2?

    _____ A. Lateral collateral ligament

    _____ B. Anterior cruciate ligament

    _____ C. Medial meniscus

    _____ D. Posterior cruciate ligament

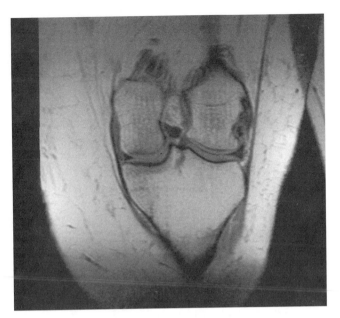 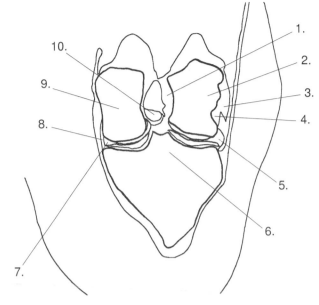

Figure 7-17

1. Which of the following is illustrated by 4?
   _____ A. Lateral collateral ligament
   _____ B. Lateral meniscus
   _____ C. Lateral epicondyle
   _____ D. Popliteal tendon

2. What number illustrates the medial meniscus?
   _____ A. 7
   _____ B. 5
   _____ C. 8
   _____ D. 10

3. Which of the following is illustrated by 10?
   _____ A. Lateral meniscus
   _____ B. Posterior cruciate ligament
   _____ C. Medial meniscus
   _____ D. Anterior cruciate ligament

4. What number illustrates the tibial intercondylar eminence?
   _____ A. 6
   _____ B. 10
   _____ C. 8
   _____ D. 5

5. Which of the following is illustrated by 2?
   _____ A. Lateral femoral epicondyle
   _____ B. Lateral femoral condyle
   _____ C. Head of fibula
   _____ D. Proximal tibia

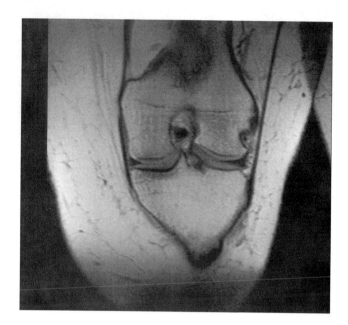 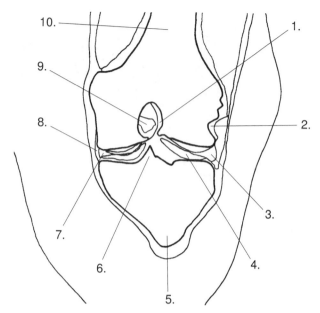

Figure 7-18

1. What number illustrates the medial collateral ligament?
   _____ A. 8
   _____ B. 7
   _____ C. 4
   _____ D. 9

2. Which of the following is illustrated by 6?
   _____ A. Articular surface of tibia
   _____ B. Intercondylar eminence
   _____ C. Medial meniscus
   _____ D. Lateral meniscus

3. Which of the following is illustrated by 9?
   _____ A. Anterior cruciate ligament
   _____ B. Intercondylar eminence
   _____ C. Posterior cruciate ligament
   _____ D. Medial collateral ligament

4. What number illustrates the popliteal tendon?
   _____ A. 1
   _____ B. 9
   _____ C. 3
   _____ D. 2

5. Which of the following is illustrated by 4?
   _____ A. Articular surface of tibia
   _____ B. Intercondylar eminence
   _____ C. Medial meniscus
   _____ D. Lateral meniscus

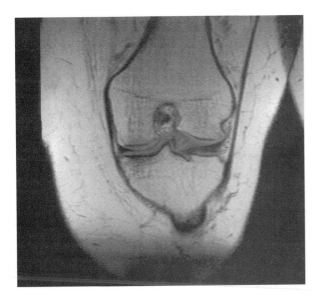 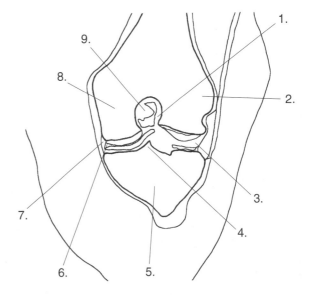

Figure 7-19

1. What number illustrates the posterior cruciate ligament?

    _____ A. 1

    _____ B. 9

    _____ C. 4

    _____ D. 3

2. Which of the following is illustrated by 7?

    _____ A. Medial meniscus

    _____ B. Intercondylar eminence

    _____ C. Medial collateral ligament

    _____ D. Lateral meniscus

3. What number illustrates the anterior cruciate ligament?

    _____ A. 1

    _____ B. 8

    _____ C. 2

    _____ D. 9

4. Which of the following is illustrated by 3?

    _____ A. Articular surface of medial condyle

    _____ B. Articular surface of tibia

    _____ C. Anterior cruciate ligament

    _____ D. Lateral meniscus

5. What number illustrates the medial meniscus?

    _____ A. 6

    _____ B. 7

    _____ C. 3

    _____ D. 9

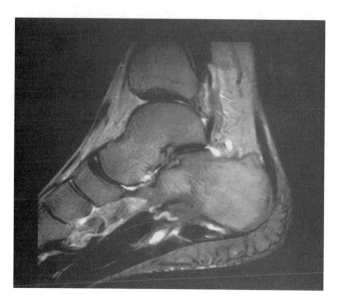
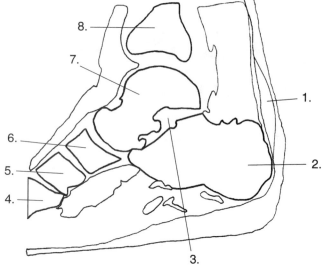

Figure 7-20

1. Which of the following is illustrated by 4?

_____ A. First cuneiform

_____ B. Navicular

_____ C. First metatarsal

_____ D. First digit

2. What number illustrates the tibia?

_____ A. 1

_____ B. 8

_____ C. 3

_____ D. 7

3. Which of the following is illustrated by 7?

_____ A. Navicular

_____ B. Calcaneus

_____ C. Tibia

_____ D. Talus

4. What number illustrates the tendinous insertion of gastrocnemius muscles?

_____ A. 1

_____ B. 3

_____ C. 6

_____ D. 2

5. Which of the following is illustrated by 5?

_____ A. First cuneiform

_____ B. Talus

_____ C. Navicular

_____ D. First metatarsal

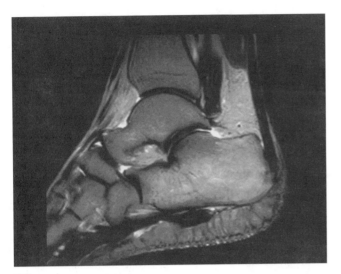 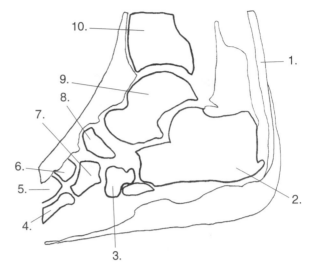

Figure 7-21

1. Which of the following is illustrated by 6?

    _____ A. Third metatarsal

    _____ B. Second metatarsal

    _____ C. Third cuneiform

    _____ D. Second cuneiform

2. What number illustrates the talus?

    _____ A. 7

    _____ B. 9

    _____ C. 8

    _____ D. 2

3. What number illustrates the third cuneiform?

    _____ A. 6

    _____ B. 4

    _____ C. 7

    _____ D. 5

4. Which of the following is illustrated by 2?

    _____ A. Calcaneus

    _____ B. Talus

    _____ C. Tibia

    _____ D. Tendocalcaneus tendon

5. Which of the following is illustrated by 8?

    _____ A. Cuboid

    _____ B. Navicular

    _____ C. Talus

    _____ D. Second cuneiform

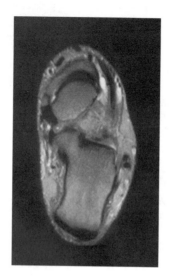

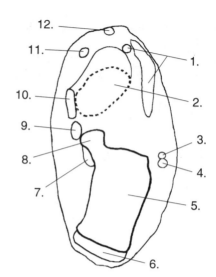

M   L

Figure 7-22

1. What number illustrates the tibialis anterior tendon?
    _____ A. 1
    _____ B. 11
    _____ C. 9
    _____ D. 12

2. Which of the following is illustrated by 8?
    _____ A. Cuboid
    _____ B. Talus
    _____ C. Navicular
    _____ D. Sustentaculum tali

3. Which of the following is illustrated by 3?
    _____ A. Peroneus longus tendon
    _____ B. Flexor hallucis longus tendon
    _____ C. Flexor digitorum longus tendon
    _____ D. Peroneus brevis tendon

4. What number illustrates the extensor digitorum tendon?
    _____ A. 1
    _____ B. 11
    _____ C. 4
    _____ D. 10

5. What number illustrates the extensor hallucis longus tendon?
    _____ A. 11
    _____ B. 1
    _____ C. 12
    _____ D. 10

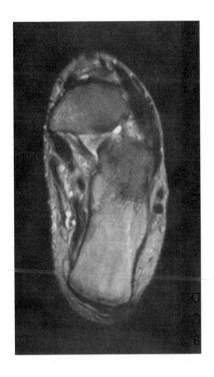

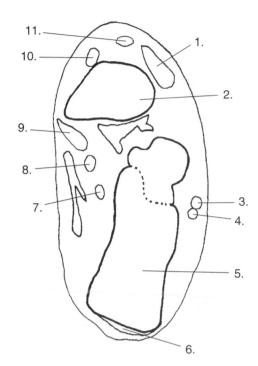

M   L

Figure 7-23

1. What number illustrates the tibialis posterior tendon?

    _____ A. 10

    _____ B. 3

    _____ C. 9

    _____ D. 7

2. Which of the following is illustrated by 10?

    _____ A. Extensor hallucis longus tendon

    _____ B. Tibialis anterior tendon

    _____ C. Extensor digitorum tendon

    _____ D. Tibialis posterior tendon

3. What number illustrates the peroneus brevis tendon?

    _____ A. 3

    _____ B. 7

    _____ C. 4

    _____ D. 8

4. Which of the following is illustrated by 5?

    _____ A. Cuboid

    _____ B. Talus

    _____ C. Sustentaculum tali

    _____ D. Calcaneus

5. What number illustrates the talus?

    _____ A. 2

    _____ B. 5

    _____ C. 1

    _____ D. 9

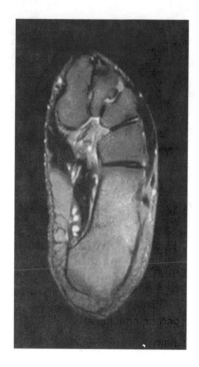

M   L

Figure 7-24

1. Which of the following is illustrated by 4?

   _____ A. Cuboid

   _____ B. Third cuneiform

   _____ C. Second cuneiform

   _____ D. Navicular

2. What number illustrates the 3rd cuneiform?

   _____ A. 2

   _____ B. 7

   _____ C. 4

   _____ D. 3

3. What number illustrates the second cuneiform?

   _____ A. 3

   _____ B. 2

   _____ C. 4

   _____ D. 1

4. Which of the following is illustrated by 5?

   _____ A. Peroneus longus tendon

   _____ B. Extensor digitorum tendon

   _____ C. Peroneus brevis tendon

   _____ D. Extensor hallucis longus tendon

5. Which of the following is illustrated by 1?

   _____ A. Second cuneiform

   _____ B. Third cuneiform

   _____ C. First cuneiform

   _____ D. Cuboid

ANSWERS

Fig. 7-1

1. A
2. B
3. D
4. A
5. C

Fig. 7-2

1. D
2. B
3. C
4. A
5. C

Fig. 7-3

1. D
2. A
3. A
4. C
5. D

Fig. 7-4

1. B
2. D
3. A
4. C
5. C

Fig. 7-5

1. B
2. D
3. C
4. A
5. A

Fig. 7-6

1. C
2. A
3. D
4. B
5. D

Fig. 7-7

1. D
2. C
3. A
4. B
5. C

Fig. 7-8

1. B
2. A
3. D
4. C
5. A

Fig. 7-9

1. B
2. D
3. C
4. A
5. D

Fig. 7-10

1. D
2. B
3. B
4. A
5. B

Fig. 7-11

1. B
2. C
3. A
4. A
5. A

Fig. 7-12

1. B
2. D
3. D
4. A
5. D

Fig. 7-13

1. A
2. D
3. D
4. B
5. A

Fig. 7-14

1. D
2. B
3. A
4. C
5. A

Fig. 7-15

1. B
2. D
3. D
4. A
5. D

Fig. 7-16

1. C
2. D
3. A
4. A
5. B

Fig. 7-17

1. D
2. A
3. B
4. A
5. B

Fig. 7-18

1. A
2. B
3. C
4. D
5. A

Fig. 7-19

1. B
2. C
3. A
4. D
5. A

Fig. 7-20

1. C
2. B
3. D
4. A
5. A

Fig. 7-21

1. D
2. B
3. C
4. A
5. B

Fig. 7-22

1. B
2. D
3. D
4. A
5. C

Fig. 7-23

1. C
2. B
3. A
4. D
5. A

Fig. 7-24

1. A
2. D
3. B
4. C
5. C